COMPANION TO GYNAECOLOGY

COMPANION TO GYNAECOLOGY

Christopher L.-H. Huang
New Hall, Cambridge

Victor G. Daniels
Christ's College, Cambridge

WKAP ARCHIEF

MTP PRESS LIMITED
a member of the KLUWER ACADEMIC PUBLISHERS GROUP
LANCASTER / BOSTON / THE HAGUE / DORDRECHT

Published in the UK and Europe by
MTP Press Limited
Falcon House
Lancaster, England

British Library Cataloguing in Publication Data

Huang, Christopher L.H.
  Companion to gynaecology.—(Companion series)
  1. Gynecology
  I. Title   II. Daniels, Victor G.
  618.1   RG101

Published in the USA by
MTP Press
A division of Kluwer Boston Inc
190 Old Derby Street
Hingham, MA 02043, USA

Library of Congress Cataloging in Publication Data

Huang, Christopher L.-H.
  Companion to gynaecology.

  Includes index.
  1. Gynecology—Outlines, syllabi, etc.
  I. Daniels, Victor G.   II. Title.   [DNLM:
  1. Gynecology—handbooks.   WP 39 H874c]
  RG112.H83.   1985   618.1   85–9996

ISBN-13: 978-94-010-8655-4   e-ISBN-13: 978-94-009-4870-9
DOI: 10.1007/978-94-009-4870-9

Photoset and printed by
Redwood Burn Limited,
Trowbridge, Wiltshire

Contents

Contents

Preface

This book was written to provide a clear and systematic summary of the principles of gynaecology in synoptic form. It is part of a three volume series, the first two volumes covering topics in undergraduate obstetrics and neonatal medicine respectively.

Where appropriate, fundamentals of related anatomy and physiology are also covered. It is primarily directed at undergraduate medical students and midwives, but material useful as reference to doctors revising for further qualifications has been included. Although much of the content is organized in the form of lists, this book differs from the usual 'list' book in that coverage is full and systematic. Thus the text has been organized into the following subheadings, where appropriate: Definitions, Aetiology, Pathophysiology, Clinical features, Differential Diagnosis, Investigations, Treatment, and Prognosis. To avoid repetition, certain clinical areas including ectopic pregnancy and abortion which have been covered in the earlier volumes are not included.

Certain useful diagnostic lists are also provided, and, in preparing the book, previous examination papers of the Universities of Oxford, Cambridge and London, as well as of the Central Midwives Board, were consulted. Illustrations have been specially prepared in the form of explanatory line drawings that are simple, and easy to memorize and reproduce. Although drug dosages were checked with care before going to press, changes in medical practice make it advisable to verify regimes and doses with the latest prescribing information and the pharmacopoeia, before use.

Cambridge
January 1985

CHRISTOPHER L.-H. HUANG
VICTOR G. DANIELS

1

Clinical Methods

HISTORY

1 Age, gravidity, parity

2 Presenting complaint with history of present illness

3 Obstetric history:
   a Previous pregnancies:
      i Dates and duration and place of birth
      ii Complications during pregnancy
      iii Character and duration of labour
      iv Outcome – weight and sex of infant
      v History of puerperal infection
   b History, if any, of stillbirths and abortions

4 General history:
   a Previous cardiovascular, renal or infectious disease
   b Past operations
   c Psychiatric history, if any

5 Gynaecological history:
   a Pain – ask for site, nature and occurrence of pain and for dyspareunia
   b Discharge – ask about sexually transmitted diseases
   c Menstrual history:
      i Menarche
      ii Rhythm of menstrual cycle. Duration of flow. Amount of blood loss. Menstrual symptoms. Irregular bleeding, if any

       iii Menopause
       iv Use of contraceptives
   d Micturition and bowel function

6 Social history:
   a Occupation
   b Travel
   c Marital or cohabitation relationship

7 Sexual history – if relevant

8 Medications – including the oral contraceptive pill

GYNAECOLOGICAL EXAMINATION

1 Breasts
   a Appearance – size, shape, symmetry, discharge, if any
   b Palpation – texture and presence of breast lumps, if any
   c Examination of the axillae

2 Abdomen
   a Appearance – shape, size, masses if any
   b Palpation:
       i Superficial – for guarding, tenderness, herniae
       ii Deep – palpation of liver, kidneys and spleen
       iii Examination for deep masses or tenderness
   c Auscultation for intestinal activity

3 Pelvis
   a External genitalia:
       i Pubic hair – pattern masculine or feminine, nits or lice, infection
       ii Skin changes
           Areas of inflammation, atrophy or hypertrophy
           Presence or absence of vaginal discharge
           Presence or absence of areas of tenderness or defined lesions
           Urethral or duct lesions
   b Speculum examination:
       i Bivalve speculum (Cusco's speculum) (Figure

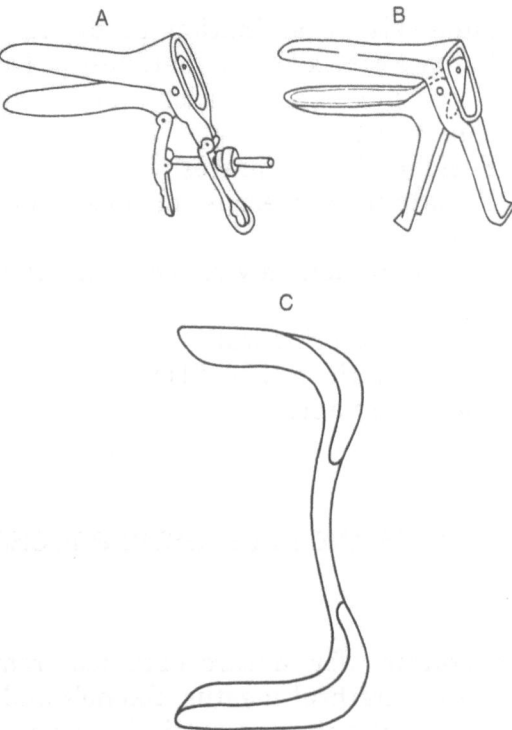

Figure 1.1 *Cusco's speculum in* (A) *steel,* (B) *plastic disposal form,* (C) *Sims' speculum*

1.1): sequential examination of labia, vagina and cervix.

ii Patient in lithotomy position

iii Cervical examination:
- Size, contour and surface features
- Presence or absence of lacerations and displacement
- External os
- Presence, if any, of discharge, blood or other fluid
- Presence of masses, if any
- Presence of tenderness on displacement

iv Duck-bill speculum (Sims' speculum) – useful for examination for prolapse or fistulae, with patient in semi-prone position

   c Bimanual pelvic examination (Figure 1.2):
- **i** Palpate cervix for any hardness or irregularity
- **ii** Note shape, size, position and mobility of uterus (Figure 1.3)
- **iii** Feel for adnexal masses
- **iv** Palpate for tenderness or masses of the bladder base

   d Rectal examination: in virgin or child. Palpation for:
- **i** Rectocoele
- **ii** Posterior uterine wall
- **iii** Masses in the pouch of Douglas
- **iv** Sacral nerve trunks

INVESTIGATIONS AND DIAGNOSTIC PROCEDURES

Microbiological:

1 Smears:
- **a** Gram-staining for smears obtained from fluid exudates from urethral meatus, Skene's and Bartholin's ducts, vaginal walls and cervical os as necessary
- **b** Separate microscopic examination of smears for *Trichomonas vaginalis*
- **c** Urine culture and microscopy from a midstream specimen of urine

2 Possible primary pathogens include:
- **a** *Neisseria gonorrhoeae*
  - **i** Smears for Gram-staining: intracellular diplococci
  - **ii** Swabs in Stuart medium from cervix, urethra and rectum
- **b** *Trichomonas vaginalis*
  - **i** Wet film examined shows motile flagellates
  - **ii** Culture – from cervical or vaginal smear in Fineberg–Whittington media
- **c** *Candida albicans*
  - **i** Stained film reveals mycelia
  - **ii** Culture from vaginal swab in Sabouraud's culture media

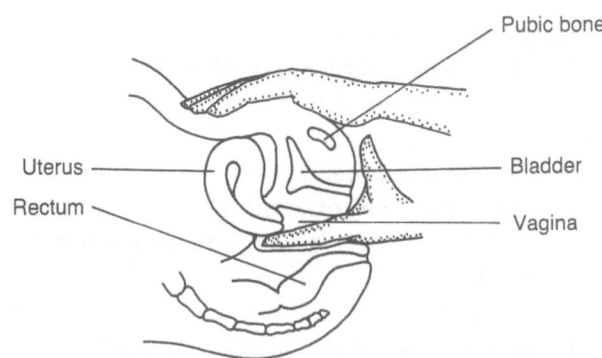

Figure 1.2   *Technique of bimanual pelvic examination*

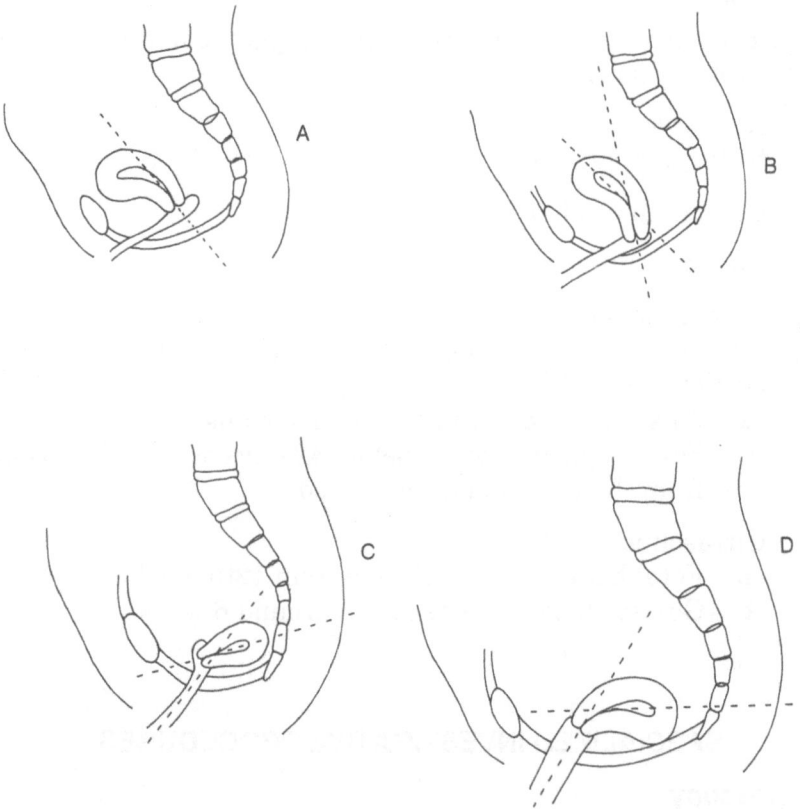

Figure 1.3   A *Anteversion of the uterus*; B *anteflexion of the uterus*;
C *retroversion of the uterus*; D *mobile retroflexion of the uterus*

    d *Treponema pallidum*
       i Wet dark field preparation from ulcer
      ii Fixed preparation for immunofluorescence test
     iii Serological tests
    e *Mycobacterium tuberculosis*
       i Premenstrual endometrium for staining for acid-fast bacilli and for culture on Lowenstein–Jensen media
    f Herpes virus
       i Smear for cytological examination – multinucleate giant cells
      ii Electron microscopy
     iii Serological tests

Cytological:
Cervical smear cytology to screen for dysplastic malignant conditions of the cervix

Biopsy:
  1 Vulva and vagina

  2 Cervix via a colposcope

  3 Endometrium via a curette

Test for pregnancy:
  1 Urinary human chorionic gonadotrophin (HCG) – pregnosticon test. NB:
    a False positives occur in the menopause
    b False negatives occur before 42 days or after 20 weeks from the last menstrual period

  2 Ultrasound (see below)
    a Fetal heart can be detected from 12th week
    b Gestation sac detected from around 6 weeks

SPECIALIZED INVESTIGATIVE PROCEDURES

Culposcopy

The culposcope is a binocular microscope used for direct visualization of exocervical, vaginal or vulvar lesions. It uses a 30–60 ×

magnification under illumination, and is especially useful in connection with diagnosis of suspected cervical carcinoma

Method:

1 Obtain cervical smear first for cytological examination

2 Cleanse cervix with 3% acetic acid to facilitate visualization

3 Focus the culposcope on the cervix, paying special attention to the squamo-columnar junction

4 When carcinoma is suspected paint on Lugol's strong iodine solution (Schiller test). Abnormal areas of epithelium do not stain

Culdoscopy

1 Principle – culdoscopy involves the inspection of pelvic structures through an incision in the vaginal vault and pouch of Douglas. The culdoscope is an optical instrument with a lens system and illumination that allows direct vision of pelvic viscera

2 Indications – presence of symptoms not explicable from history and physical findings but in which laparotomy not clearly indicated, useful in diagnosis of:
   a Ectopic pregnancy
   b Endometriosis
   c Pelvic pain of uncertain origin
   d Infertility

3 Complications:
   a Perforation of a viscus
   b Intraperitoneal bleeding
   c Peritoneal infection

Laparoscopy

1 Principle: laparoscopy is an endoscopic technique that allows inspection of the pelvic cavity. The endoscope is inserted through the abdominal wall into the peritoneal cavity

2 Indications:
- **a** Diagnostic: to examine or assess:
  - **i** Uterine and ovarian congenital abnormalities
  - **ii** Endometriosis
  - **iii** Salpingitis
  - **iv** Ectopic pregnancy
  - **v** Tubal patency
- **b** Biopsy of:
  - **i** Ovarian tumours
  - **ii** Omentum
  - **iii** Spleen
  - **iv** Liver (neoplasia, cirrhosis)
- **c** Therapeutic:
  - **i** Tubal sterilization
  - **ii** Removal of extruded intrauterine device
  - **iii** Lysis of adhesions
  - **iv** Fulguration of endometriosis

3 Contraindications:
- **a** Previous abdominal operations or peritonitis: risk of adhesions to the surgical scar
- **b** Danger of perforating a viscus
- **c** Pregnant uterus
- **d** Severe cardiac or pulmonary disease
- **e** Gross obesity

4 Method:
- **a** Anaesthetized patient
- **b** Empty bladder
- **c** Cannula and forceps fixed to cervix
- **d** 4 litres of CO_2 injected through a thin cannula in the abdominal cavity
- **e** Endoscope then passed through cannula to enable visualization of pelvic viscera
- **f** Second cannula can be introduced for purpose of introducing:
  - **i** Diathermy forceps for tubal sterilizations
  - **ii** Biopsy forceps

5 Risks:
- **a** Perforation of viscera

b Damage to major blood vessels close to sacral promontory

c Infection

Hysteroscopy

1 Principle: visualization of uterine cavity and lumens of the oviducts with a fine-bore fibreoptic endoscope

2 Indications:
  a Removal of foreign body e.g. intrauterine device
  b Diagnosis of abnormal uterine bleeding
  c Biopsy of endometrium
  d Investigation of infertility – patency of Fallopian tubes

3 Contraindications
  a Pregnancy
  b Acute cervicitis
  c Haemorrhage

4 Technique
  a Paracervical or light general anaesthesia
  b Slight distension of the uterus with dextran solution
  c Visualization by fibreoptic endoscope inserted through the vagina

Ultrasound

1 Principle: measurement of reflections of a pulsed ultrasonic beam reflected from interfaces between tissues that possess different densities

2 Indications:
  a Diagnosis of pelvic tumour
  b Differentiation of solid from cystic neoplasms
  c Diagnosis of ascites

DILATATION AND CURETTAGE

1 Procedure:
  a Dilation of the cervix using a dilator

    b Removal of the uterine contents with ovum forceps and then a curette to scrape endometrium

2 Risks:
  a Splitting the cervix during dilation
  b Perforation of the uterus

3 Indications:
  a Diagnostic indications occur in the investigation of:
   i Dysfunctional uterine bleeding
   ii Sterility
   iii Amenorrhoea
   iv Pelvic tuberculosis
   v Malignant disease of the uterus
  b Therapeutic indications:
   i Retained products of conception after abortion
   ii Termination of pregnancy
   iii Removal of polyps
   iv Dysmenorrhoea
   v Insertion of radium implants

2

Disorders of Development

INTRODUCTION

Development problems may occur at any of the following stages of development of the female genitalia:

1   Sex determination: Sex is determined by the complement of sex chromosomes from the male and female gametes. All human somatic cells contain 46 chromosomes: 22 homologous pairs and 2 sex chromosomes. It is the chromosomes that determine the sex of an individual. The normal female has two X (XX) and the normal male an X and a Y (XY) chromosome. The sex chromosome complement is determined at fertilization.

   a   Before fertilization, male and female germ cells undergo meiosis. This reduces the original set of 22 autosomes and 2 sex chromosomes to half that number

   b   Hence the (male) spermatozoon carries either an X *or* a Y chromosome, whereas the (female) oocyte always contains an X chromosome

   c   Fertilization by a spermatozoon containing a Y chromosome produces a male (XY) zygote; fertilization by a spermatozoon containing an X chromosome produces a female (XX) zygote

2   Differentiation of the gonads: In both sexes, the presumptive gonads begin as a blastema on each side of the embryo at 5–6 weeks after conception at the ventral surface. The

19

direction of differentiation of the gonads to the male or female type depends on the presence or absence of a Y chromosome. Its presence initiates development of the testis, its absence permits female development

 a In the male:
  i Sex cords form in the medulla. They give rise to the seminiferous tubules
  ii The sex cords are joined by the rete testis to the mesonephric tubules, which eventually form the epididymis

 b In the female:
  i Cells of the cortical part of the primordial gonad proliferate to form a mass of germ cells
  ii The primitive oocytes so formed are surrounded by a layer of cells derived from the epithelium covering the ovary

 c Differentiation of the internal and external genitalia depends upon differentiation of the gonad and the hormones produced therefrom:
  i Differentiation of the gonads into a testis results in production of hormones that lead to development of a male genital tract
  ii Absence of the testis and the hormones produced thereof permits development of a female genital tract

3 Development of the full normal anatomical structure of the internal and external genitalia

4 Development of full secondary sexual characteristics at puberty

5 Psychological factors: The psychological sex orientation of the individual is additionally determined by such factors as social upbringing and assigned gender identity

DIAGNOSIS OF CHROMOSOMAL ABNORMALITIES

1 Autosomal studies:
 a Cultured lymphocytes are transformed *in vitro* into lymphoblasts by the addition of phytohaemagglutinin

b The resulting mitosis of the lymphoblasts is arrested at metaphase by adding colchicine

c The chromosomes are then dispersed for microscopic study using hypotonic solutions. They are grouped as pairs 1–22 according to size into groups A to G

2 Detection of the X chromosome: Testing is normally done in buccal smear cells.

    a In female cells one of the two X chromosomes can be detected as the Barr body in 15–20% of interphase cells

    b The Barr body is absent in normal male cells – this is because the number of Barr bodies is always one less than the total number of X chromosomes

3 Detection of the Y chromosome. This is done in buccal smear cells by staining with the fluorescent dye, quinacrine hydrochloride

DISORDERS OF CHROMOSOMAL SEX DETERMINATION

They may occur during the formation of germ cells or during fertilization.

1 Sex chromosome aberrations during germ cell formation

    a Possible problems: Errors in division of the originally paired sex chromosome during meiosis may result from:

        i Meiotic non-disjunction – homologous chromosomes fail to separate during the anaphase of the first maturation division

        ii Anaphase lag – a chromosome is lost as the chromosomes migrate to opposite poles during anaphase

    b Consequences: The errors will result in a gamete with an extra chromosome or a gamete deficient in one chromosome. The zygote that results from subsequent fertilization will then contain three ('trisomy') or only one ('monosomy') of the chromosomes affected. The possible permutations are summarized in Table 2.1.

Table 2.1   Sex chromosomal complements in the zygote consequent upon errors during germ cell formation

Male gamete (spermatozoon)	Female gamete (ovum)		
	X	XX	O
X	XX normal female	XXX triple X superfemale	XO Turner's syndrome
Y	XY normal male	XXY Klinefelter's syndrome	YO lethal
XY	XXY Klinefelter's syndrome	XXXY male triple X; variation of Klinefelter's syndrome	XY normal male
O	XO Turner's syndrome	XX normal female	OO lethal

2 Sex chromosomal aberrations during or after fertilization. Problems identical to those described occurring before fertilization can occur, but additional problems could arise from:
   a Mitotic non-disjunction
   b Chromosome loss during mitotic anaphase
These may occur either during the first mitotic division after fertilization or one of the early zygotic divisions that follow. The result is a chromosomal mosaic.

DISORDERS OF GONADAL DEVELOPMENT

In addition to sex chromosomal aberrations, these may result from genetic mutations, or from extraneous factors such as irradiation or infection.

Gonadal agenesis

   1 Basic cause: The gonadal anlage is missing or damaged. As a result, the genital tract and external genitalia are female. The condition is rare

2 Clinical features:
   a Persistent infantile habitus
   b Primary amenorrhoea
   c Severe genital hypoplasia
   d Lack of secondary sexual characteristics
   e Psychically – female

3 Investigations:
   a Chromosomes – normal male or female
   b Diagnosis confirmed by laparoscopy or laparotomy

4 Management:
   a Patient may need to be informed about infertility
   b Females may require long-term oestrogen therapy

Gonadal dysgenesis

This is the arrest of development of the gonads before or immediately after gonadal differentiation. Hence neither ovarian nor testicular tissue is present.
   These are characterized by:
1 Imperfectly formed gonads containing no germ cells
2 Immature female habitus
3 Infantile vulva, vagina and uterus

Turner's syndrome (Figure 2.1)

This defines a variety of gonadal dysgeneses with short stature and other anomalies.

1 Genetics:
   a Usually – 45 chromosomes, XO genotype with consequent arrest of gonadal differentiations. However, variants on this include:
   b Local mosaic patterns involving an XX cell line that will result in structures that resemble primordial follicles

2 Pathophysiology: the abnormal sex chromosome complement results in early arrest of gonadal development germ cells and follicular cords do occur in early stages of development, but degeneration of the follicular cells and failure of

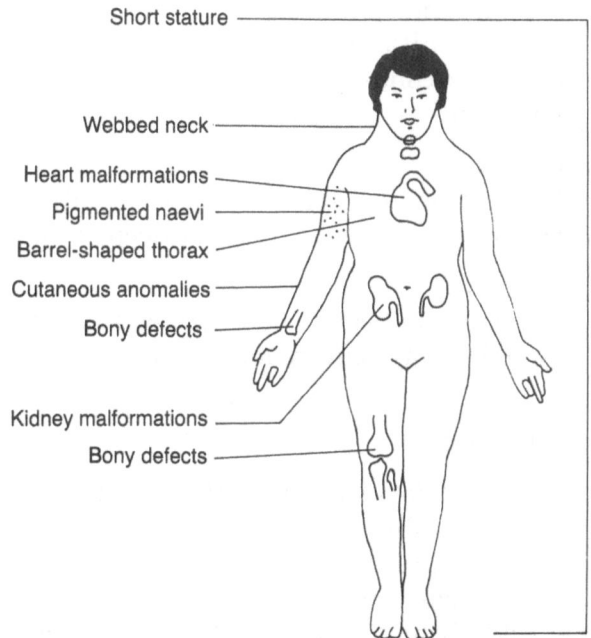

Short stature
Webbed neck
Heart malformations
Pigmented naevi
Barrel-shaped thorax
Cutaneous anomalies
Bony defects
Kidney malformations
Bony defects

Figure 2.1   *Clinical features of Turner's syndrome*

formation of primordial follicles subsequently occurs. The degenerating structures are replaced by connective tissue. Most of the embryos that are formed with such a chromosome complement in fact abort at an early stage of development

3   Clinical features:
  a   Short stature
  b   Webbed neck; 'sphinx' face and 'carp' mouth
  c   Shield-shaped thorax; widely spaced nipples; poor breast development
  d   Coarctation of the aorta
  e   Cubitus valgus (increased carrying angle of the upper limb); short fourth metacarpal; fingernail hypoplasia
  f   Sterile gonads; primary amenorrhoea and infertility

g Multiple pigmented naevi
h Kidney and ureteric deformities

4 Differential diagnosis:
    a Ovarian hypoplasia
    b Delayed puberty
    c Pituitary dwarfism

5 Investigations:
    a Recognition of clinical features
    b Chromosome analysis
    c Elevated serum levels of follicle-stimulating hormone and luteinizing hormone
    d Elevated urinary gonadotrophins
    e Reduced urinary oestrogens and 17α-ketosteroids
    f Laparoscopy

6 Management:
    a Counselling
    b Oestrogen and progesterone therapy to initiate feminization, beginning at around 12–13 years

'Pure' gonadal dysgenesis

This must be distinguished from gonadal dysgenesis found in Turner's syndrome.

1 Genetics: basic cause is unknown – where the chromosomal complement has been studied, a varying sex chromosomal complement has been demonstrated. It may therefore be the result of loss of genes not detected by standard cytogenetic methods, or may even not be of genetic origin

2 Pathophysiology: Whatever the cause, 'pure' gonadal dysgenesis manifests exclusively on interference with gonadal development

3 Clinical features:
    a Development of genitalia along female lines, and the gonadal streaks found in Turner's syndrome do occur
    b Stigmata described for Turner's syndrome are absent
    c Phenotypically – females with normal stature
    d At puberty, internal and external genitalia remain in-

fantile; patients often present with primary amenorrhoea

4 Diagnosis: as for Turner's syndrome – definitive diagnosis
by laparotomy or laparoscopy

Ovarian or testicular dysgenesis

This describes arrest of gonadal development, but after differentiation into specific male or female structures, and its extent is
therefore dependent upon both the timing and the extent of this
interruption.

Ovarian dysgenesis:

1 Genetics: the sex chromosome complement is that of a
normal female (XY)

2 Pathophysiology:
    a Afollicular type – primordial follicles absent, although
       other structures typical of normal ovary may be pres-
       ent
    b Follicular type – primordial follicles do occur in the
       inner layers of the ovarian cortex, but mature follicles
       are absent

3 Clinical features:
    a Primary amenorrhoea
    b Female phenotype and gender role
    c Secondary sexual characteristics only moderately de-
       veloped

4 Diagnosis: ovarian biopsy is needed to diagnose definitively

5 Management:
    a Amenorrhoea can be corrected by hormone therapy
    b Oestrogen substitution therapy is needed. However,
       infertility cannot be corrected as ovulation may not be
       inducible

Triple X syndrome:

1 Genetics: presence of a triple complement of X chromosome
with normal autosomal complement

2   Pathophysiology: a varying degree of ovarian dysgenesis may be present but some patients may have normal ovarian function and fertility

3   Clinical features:
   a   Higher incidence of mental retardation
   b   Phenotype otherwise unremarkable
   c   Some cases present with amenorrhoea and infertility

Testicular dysgenesis:

Failure of testicular development may occur at any stage after differentiation begins.

1   An arrest at an early stage may cause a failure of hormone production by the testis. This will result in the embryo developing along female lines, and an intersexual individual

2   If the testes cease to produce androgen at a later stage of development, isolated clinical stigmata may occur: the best known example is Klinefelter's syndrome

Testicular dysgenesis due to Klinefelter's syndrome

Incidence: 1 = 400 newborns

1   Pathophysiology:
   a   Chromosomal complement is 47 autosomes and XXY sex chromosomes. This results from a non-disjunction resulting in the maternal zygote receiving an extra X chromosome
   b   In the subject there is an initial normal development of testes and secondary sexual characteristics

2   Clinical features:
   a   Infantile genitalia
   b   Eunuchoid habitus with possibly gynaecomastia and features of feminization
   c   Often tall stature

3   Diagnosis: analysis of Barr bodies and karyotypes

Testicular dysgenesis with intersexuality

1   Pathophysiology: A failure occurs in formation of the gonads at the time of differentiation of internal and exter-

nal genitalia. There may then be a failure of production of androgens, or of 'factor X'

2 Clinical features: The overall appearance and degree of development in the male or female direction determined by the time and degree of gonadal failure
   a Failure of synthesis of factor X will result in the production of both Wolffian and Mullerian ducts.
   b Failure of synthesis of androgens results in failure of development of both Wolffian and Mullerian ducts.

Testicular dysgenesis with partial feminization

1 Pathophysiology: The chromosomal pattern is 46 autosomes and XY, but the testes are defective and the genital tracts show various combinations of male and female features

2 Clinical features: external genitalia may be predominantly male, ambivalent or female

3 Management: often there is a problem of clarification as male or female. Often it is desirable to attach a female gender identity during psychological management. With later development surgical procedures may be necessary to correct deformities of apparently the opposite sex

Testicular dysgenesis with total feminization ('testicular feminization syndrome')

1 Pathophysiology: chromosomal sex is 46, XY yet gonadal sex is entirely female

2 Clinical features:
   a Patient is phenotypically female
   b External genitalia are female; clitoris often small
   c Internal genitalia – vagina may only be a blind pouch, with missing uterus and Fallopian tubes
   d Pubic and axillary hair greatly reduced
   e Testes may occur in the inguinal canal or labia majora; there is a high incidence of testicular neoplasia

3 Investigation:
   a Chromosomal pattern 46/XY

    b Normal steroid levels; elevated gonadotrophin levels

4 Management:
    a Orchidectomy – because of frequent malignant change
    b Give long-term oestrogen therapy
    c Surgical enlargement may be required for the vagina
       if hypoplastic

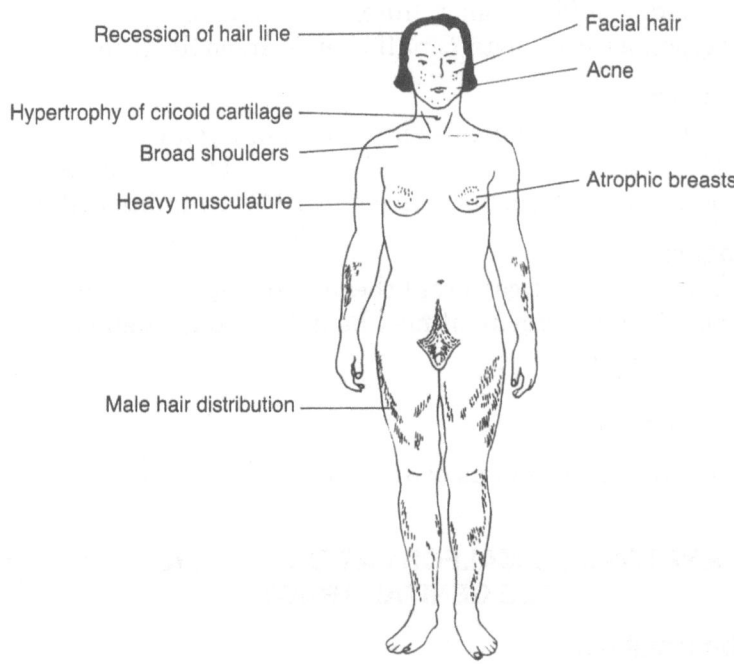

Figure 2.2 *Clinical features of virilization in the adrenogenital syndrome*

Adrenogenital syndrome (Figure 2.2)

1 Possible causes:
    a Deficiency in activity of enzymes synthesizing cortisol.
       This may occur as a recessive hereditary disorder with
       an incidence of 1:5000 newborns. Cortisol deficiency
       results in excess adrenocorticotrophic hormone
       (ACTH) secretion. ACTH causes adrenal hyperplasia
       and excess production of steroid precursors and sex
       hormones

    b Postpubertal presentation may result from adrenal hyperplasia, adenoma, or, rarely, carcinoma

2 Clinical features:
   a Virilization of external genitalia, with even scrotum-like labia which may result in incorrect assignment of sex
   b Initial rapid growth subsequently stunted by premature epiphyseal fusion – resulting in short stature
   c Broad shoulders, short limbs, android pelvis
   d Hypoplastic breasts. Small uterus, hypoplastic ovaries

3 Investigation:
   a Chromosome studies – chromosomal female
   b Elevated 24 h urinary 17-ketosteroid and pregnanetriol excretion, normalized by ACTH administration

4 Management:
   a Long-term corticosteroid therapy in congenital cases
   b Surgical management may be indicated in adenoma or carcinoma

True hermaphroditism

Patients have testes as well as ovaries. Extremely rare.

ANATOMICAL MALFORMATIONS OF THE FEMALE GENITAL TRACT

Uterine abnormalities

1 Pathophysiology: These may result from:
   a Agenesis or inadequate differentiation of the uterine part in one (or both) Mullerian ducts.
   b Anomalous fusion of the Mullerian ducts.
   c Failure of full involution of the septum after the Mullerian ducts have fused.
   d Retarded growth of an otherwise normally differentiated uterus.

2 Uterine aplasia: The uterus is entirely absent, and replaced by a cord of connective tissue. This is the result of failure of differentiation of the uterine parts of both Mullerian ducts

3 Uterus unicornis: This results from aplasia of the uterine part of one Mullerian duct. One uterine horn only is present

4 Bicornuate uterus: A partial failure of fusion of the Mullerian ducts (in the lower portion) results in a uterus with two horns, but one cervix and vagina

5 Separate uterus: The Mullerian ducts are completely fused, but there is a persistent midline septum

6 Infantile uterus: The uterus remains hypoplastic and of prepubertal proportions even in adulthood: this may reflect a lack of sensitivity to hormones

7 Clinical features: One or more of the following:
   a During the reproductive age:
      i Infertility
      ii Habitual abortion
      iii Increased incidence of unusual or unstable fetal lie
      iv Cervical dystocia
      v Postpartum haemorrhage
   b Before the reproductive age:
      i Primary amenorrhoea
      ii Primary dysmenorrhoea in a bicornuate uterus with non-communicating horn

8 Management: Corrective surgery is indicated only if symptomatic or if there is habitual abortion

Vaginal malformations

1 Pathology: May either occur alone, or be associated with anomalies of the uterus and/or the urinary tract.
   Examples:
   a Aplasia of the vagina – usually associated with a rudimentary uterus. However, the ovaries are usually functionally and structurally normal
   b Atresia of the vagina – this is the result of failure of canalization of the vaginal cord, from which the vagina develops
   c Septate vagina – (often associated with double uterus)

this is the result of a persisting central part of the vagina

2 Clinical features:
   a Primary amenorrhoea
   b Haematocolpos; haematometra in partial atresis
   c Dyspareunia – due to midline septum

3 Management: Constructive plastic surgery may be indicated

Atresia of the hymen

1 Pathophysiology: Failure of penetration of the caudal part of the developing vagina into the urogenital sinus

2 Clinical features: Retention of blood at menstruation beginning at menarche results in:
   a Lower abdominal pain
   b Absence of external bleeding
   c Development of haematocolpos, or haematometra. Bluish bulging membrane on vaginal examination; cystic mass on rectal examination.

3 Management: Incision of the hymen and drainage

PATHOLOGY OF PUBERTY

Precocious puberty

Incidence: 1:20 000 girls; 1:4000 boys. Note that there is a wide variation in the time of onset of 'normal' puberty.

1 Pathophysiology:
   a True precocious puberty: this may reflect a central nervous system disturbance, e.g. brain tumour, hydrocephalus, hypothalamic lesions. A normal puberty occurs prematurely, with the expected associated endocrine alterations. Albright's disease is a rare form, it has an associated fibrotic bone dysplasia and skin pigmentation.
   b Pseudopubertal precosity: this is the result of hormone-producing tumours, for example, of the

adrenal or ovary. The clinical course of the condition is determined by the hormone produced.

2 Clinical features:
   a Premature onset of breast development and pubic hair
   b Subsequent onset of menstrual bleeding
   c Early height gain – but this subsequently is arrested by premature fusion of the epiphyses

3 Investigation:
   a History and clinical examination
   b Radiological studies to assess bone age
   c Skull X-rays – in particular to view the sella turcica
   d Intravenous pyelogram
   e Vaginal cytology
   f Endocrine investigations:
      i Gonadotrophins
      ii 17-ketosteroids
      iii Testosterone
      iv Pregnanediole
   g Laparoscopy – to check for ovarian neoplasms
   h Pneumoperitoneum – to check adrenal glands

4 Management: Treat underlying cause. Albright's disease is managed expectantly; there is a good prognosis

Delayed puberty

1 Pathology: Generally no lesion is found, parenchymal defects of the ovary or genetic factors may be involved

2 Management: Gonadal dysgenesis and lesions of the thalamus/hypothalamus should be excluded. If no organic lesions occur the patient should be reassured that sexual development will eventually be normal.

3

Menstrual Abnormalities

THE PREMENSTRUAL SYNDROME

Definition: This is a recurrent (monthly) disorder characterized by physical and psychological symptoms. The more common physical symptoms are related to water retention. The psychological changes include depression, anxiety and irritability.

Incidence: Up to 50% of women

Risk groups:
1 30–40 years of age

2 Emotional, unmarried or nulliparous women

Pathophysiology:
1 Occurs during the luteal phase of the menstrual cycle

2 The renin–angiotensin system and aldosterone along with prolactin have been implicated

3 Psychological factors may also play a part – premenstrual symptoms occur more often in neurotic patients

Clinical features:
1 Typical physical symptoms:
   a Weight increase
   b Bloatedness
   c Peripheral oedema

    d Breast swelling and pains
    e Headache or migraine
    f Backache
    g Change in bowel habit
    h Change in micturition
    i Skin changes
    j Rhinitis
    k Palpitations

2 Typical psychological symptoms:
    a Depression and disturbance of sleep
    b Anxiety
    c Nervousness and irritability
    d Aggression
    e Lethargy
    f Change in libido
    g Loss of concentration
    h Clumsiness

Management:
1 Simple explanation and sympathy

2 Diuretics, e.g. thiazides, for last 10 days of the menstrual cycle

3 Mild tranquillizers if marked mood changes

4 Treatment using the prolactin inhibitor bromocryptine – reduces breast symptoms only

5 Progesterone may be used where there is evidence of progesterone deficiency. The oral contraceptive improves symptoms in many patients

AMENORRHOEA

Terms:
1 Amenorrhoea: absence of menstruation for longer than twice the duration of a normal menstrual cycle

2 Primary amenorrhoea: non-appearance of menstrual periods in girl over the age of sixteen

3 Secondary amenorrhoea: refers to the cessation of mensis in a woman who has previously menstruated regularly

Aetiology

Physiological amenorrhoea:
1 Before menarche, preceding cyclic gonadotrophin hormone release by the hypothalamus

2 Pregnancy: Amenorrhoea persists due to elevated levels of oestrogen and progesterone (commonest causes of secondary amenorrhoea)

3 Lactation: Ovulation is suppressed by prolactin

4 Menopausal: Ovarian failure to respond to gonadotrophins

Pathological amenorrhoea:
1 General causes: Infections, malignancy, general debilitation

2 Local causes:
   a Genital tract:
      i Cryptomenorrhoea due to imperforate hymen, vaginal atresia or cervical stenosis
      ii Uterine congenital abnormalities or absence
      iii Pelvic disease, including tuberculosis
   b Ovarian:
      i Hermaphroditism – presence of both testes and ovaries
      ii Pseudohermaphroditism – where female genitalia co-exist with male gonads
      iii Gonadal dysgenesis – failure of primitive germ cells to migrate to the ovary during development, resulting in failure of ovarian function
      iv Turner's syndrome (XO genotype)
   c Iatrogenic:
      i Oophorectomy
      ii Irradiation
      iii Hysterectomy

3 Endocrine causes:
   a Cerebral:
      i Emotional upsets and stresses
      ii Anorexia nervosa
      iii Psychoses
   b Hypothalamic:
      i Iatrogenic – contraceptive ingestion, or 'post-pill' amenorrhoea. Phenothiazines and reserpine
      ii Primary hypothalamic failure – Frohlich's syndrome:
   - Begins in childhood
   - Genital hypoplasia
   - Amenorrhoea
   - Obesity
   - Somnolence
   c Pituitary:
      i Hyperprolactinaemia: Is the cause of amenorrhoea in almost one in five cases.
   - Pituitary prolactin depresses the release of gonadotrophic releasing factors and depresses the LH surge. It may also depress the sensitivity of the ovaries, preventing production of the ovarian follicle
   - Chiari–Frommel syndrome – postpartum persistence of amenorrhoea and galactorrhoea
   - Forbes–Albright syndrome – amenorrhoea and galactorrhoea due to a pituitary tumour
   - Ahumada–del Castillo syndrome – amenorrhoea, anovulation and galactorrhoea in the absence of pregnancy or pituitary tumour
      ii Acromegaly (adenoma)
      iii Hypopituitarism, for example Sheehan's syndrome. This is due to ischaemic necrosis of the pituitary following shock associated with childbirth
   d Anovulation:
      i Deficient oestrogen peak resulting in failure of the luteinizing hormone peak – this causes failure of ovulation and continued endometrial prolifer-

ation and may result in considerable break-through bleeding

ii Neoplasms:

- Androgen-producing ovarian tumours (Ar-rhenoblastomas) inhibit hypothalamic sites regulating the menstrual cycle
- Oestrogen-producing tumours. Both of these are very rare

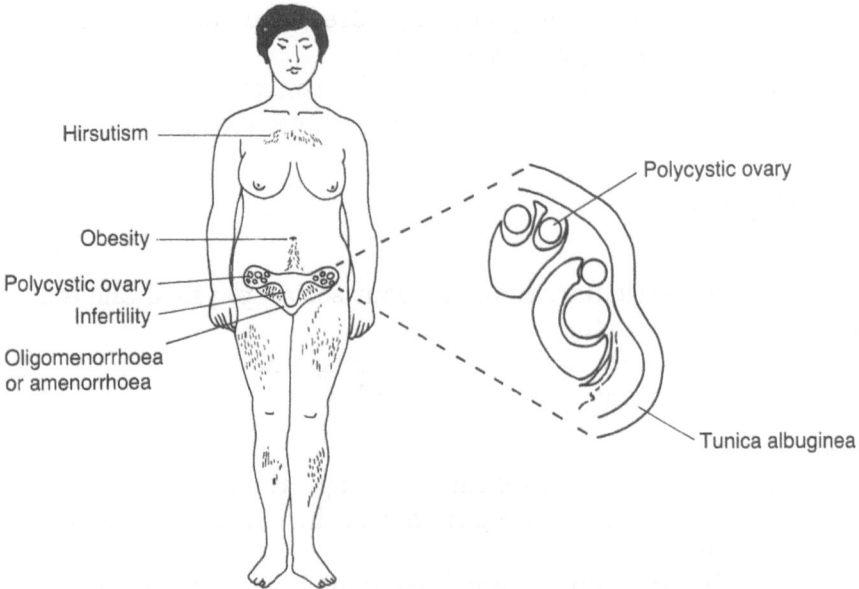

Figure 3.1   *Stein–Leventhal syndrome*

iii Stein–Leventhal syndrome (Figure 3.1): There is an ovarian enzyme defect (19-hydroxylase) in the biosynthesis of oestradiol from steroid precursors. As a result androgenic hormones are produced with the following clinical picture:

- Amenorrhoea
- Obesity
- Hirsutism
- Infertility
- Bilateral polycystic ovaries

e Adrenal cortex:

　　i　Congenital adrenal hyperplasia (Adrenogenital syndrome): This is due to enzyme defects (90% cases 21-hydroxylase, 10% cases 11-hydroxylase) in the biosynthetic pathway from progesterone to cortisol. Cortisol levels thus fall and this produces increased adrenocorticotrophic hormone (ACTH) production. Adrenal androgen production thus increases and causes masculinization of female genitalia

　　ii　Androgen-producing adrenal tumours – rare

f　Thyroid: hyperthyroidism

g　Diabetus mellitus if poorly controlled – the well-controlled diabetic has normal menstruation

Investigation

1　History:
　　a　In particular, enquire about social, physical and emotional factors
　　b　Consider the possibility of pregnancy
　　c　Ask about previous weight loss

2　Examination:
　　a　Full physical examination, in particular consider presence of recent weight change, hirsutism and galactorrhoea
　　b　Pelvic examination: in particular examine for congenital genital tract abnormalities, such as vaginal atresia

3　Special investigation:
　　a　Pregnancy should be excluded at an early stage
　　b　Plasma and urinary hormone assays:
　　　　i　Oestrogens
　　　　ii　Prolactin, follicle stimulating hormone (FSH) and luteinizing hormone levels (LH)
　　　　iii　Steroids
　　c　Chromosomal studies – buccal or vaginal smear
　　d　Vaginal cytology and endometrial biopsy to assess hormonal status
　　e　Radiological studies:
　　　　i　X-rays of the sella turcica to investigate atrophy

of the clinoid process in pituitary tumours

  ii Intravenous pyelography or retroperitoneal air insufflation studies to detect adrenal neoplasms

  iii Hysterosalpingography to detect uterine abnormalities

f Laparoscopy or culdoscopy to determine structural ovarian abnormalities

g Neurological – visual field examination to find constriction that would result from a pituitary tumour. Also, examine the fundus

Management

1 Aim to initiate spontaneous ovulatory cycles and their continuation after treatment

2 Manage the underlying cause paying adequate attention to emotional problems, diet and adequate rest

Examples:

  a Hyperprolactinaemia is treated by bromocryptine (2.5 mg two or three times daily) which inhibits prolactin secretion by the anterior pituitary

  b Stein–Leventhal syndrome is treated by clomiphene (50 mg daily for 5 days) or by wedge resection of the ovaries

3 Hormone therapy using human menopausal gonadotrophin (HMG) or human pituitary gonadotrophin to induce ovulation

Prognosis: Success in restoring periods is more difficult the longer the duration of amenorrhoea

DYSMENORRHOEA

Dysmenorrhoea refers to pain just before and during menstruation. It is the commonest of all gynaecological complaints and is the main gynaecological cause of absenteeism from work and school.

Types of dysmenorrhoea:

1   Primary or spasmodic dysmenorrhoea

2   Secondary or congestive dysmenorrhoea

Primary dysmenorrhoea

Definition: Pain with menstruation for which no organic basis is evident – intrinsic or idiopathic dysmenorrhoea
Incidence: Around 70% of cases of all painful menses
Risk groups:

1   Single or infertile adolescent women

2   Sedentary, tense and introspective patients

Pathophysiology: Uncertain – usually no physical signs

1   Hormonal factor is suggested by fact that pain does not occur in anovulatory cycles

2   Psychological factors also play a part

3   Strong, uncoordinated uterine contractions and elevated uterine tone have been discerned in some patients

Clinical features

History:

1   Severe colicky (intermittent, sharp and cramping) pain in pelvis or lower back

2   Occurs early in the period with menstrual flow and is re-lieved after 1–2 days

3   Often accompanied by premenstrual tension. Freedom from pain till onset of next period

Examination:

1   Tense, anxious patient

2   Pelvic examination normal

Management:

1   Reassurance: encourage normal activities and adequate rest

2 Mild analgesia: acetylsalicylic acid 600 mg t.d.s. or other inhibitors of prostaglandins (e.g. mefenamic acid) over the painful period

3 Hormone therapy: oestrogen–progesterone therapy to suppress ovulation

4 Anticholinergic drugs and smooth muscle relaxants

5 Psychiatric management if indicated

6 Surgical treatment – to be considered only if failure of all other measures:
   a Cervical dilatation
   b Paracervical neurectomy – dissection of the sensory sympathetic and parasympathetic nerve supply to uterus

Prognosis: Usually good

Secondary dysmenorrhoea

Definition: Painful menstruation usually in association with demonstrable pelvic pathology
Pathophysiology:
   1 Retroversion and uterine malposition may follow childbirth

   2 Chronic salpingo-oophoritis, due to:
      a Puerperal infections
      b Infected abortion
      c Venereal disease
      d Appendicitis
      e Diverticulitis
      f Tuberculosis

   3 Endometriosis

   4 Adenomyosis

   5 Uterine fibroids

   6 Intrauterine or intracervical polyps

   7 Cervical stenosis

Clinical features

History:

1 Onset of pain for a few days prior to menstrual flow. May be part of the premenstrual syndrome

2 Pain aggravated during period and only gradually relieved after flow

Examination:

1 Retroversion may be present

2 In chronic salpingo-oophoritis there may be restricted mobility and tenderness of the pelvic organs

Investigations: If indicated laparoscopy is a useful diagnostic procedure

Management:

1 Relief of pain – as above
2 Treatment of the underlying condition: for example:
   a Chronic salpingo-oophoritis (see Chapter 4)
      i Short wave diathermy
      ii Antibiotics – in acute phases
      iii Bed rest – in acute phases
      iv Surgical hysterectomy may be needed in refractory cases
   b Endometriosis: Surgical intervention with removal of the ectopic tissue

ABNORMAL UTERINE BLEEDING

Definition: Any bleeding from the uterus that significantly differs from that of the menstrual cycle. This refers to both:

1 Menstrual disturbances – uterine bleeding:
   a In excess of the normal period
   b Prolonged beyond 7 days at the expected time of the menses
   c More often than at intervals of 24 days

2 Non-menstrual: irregular flow at times other than the menstrual period

Patterns of abnormal uterine bleeding:

1 Polymenorrhoea: Cyclic bleeding over less than 24 days duration may relate to a dysrhythmic release of gonado-trophins

2 Menorrhagia (hypermenorrhoea): Excessive bleeding during a menstrual cycle of normal duration. May reflect an abnormal balance in the levels of circulating oestrogen and progesterone

3 Hypomenorrhoea (cryptomenorrhoea): An abnormally small amount of menstrual flow. A hymenal or cervical obstruction may be responsible

4 Polymenorrhagia: Excessive bleeding with a menstrual cycle of reduced duration. May occur in pelvic inflammatory disease or psychosomatic disorders

5 Metrorrhagia: Excessive, prolonged bleeding irregularly and acyclically. May occur in pathological conditions of the uterus

Causes of abnormal uterine bleeding:

1 Iatrogenic:
   a Intrauterine contraceptive device
   b Neglected vaginal pessaries
   c Treatment with ovarian hormones – 'breakthrough' bleeding may occur with synthetic progestagens given for oral contraception

2 Abnormal pregnancy:
   a Threatened or incomplete abortion
   b Ectopic pregnancy
   c Hydatidiform mole

3 Systemic causes:
   a Coagulation defects or excessive capillary fragility
   b Hyper- or hypothyroidism
   c Psychosomatic causes – emotional shock
   d Hormone-secreting tumours of the ovary

4 Pelvic causes:
   a Low-grade endometritis

    b   Fibroids – cause gradually progressive menorrhagia
    c   Endometriosis
    d   Adenomyosis
    e   Uterine or cervical polyps
    f   Ulcerating cervical or uterine carcinoma
    g   Congenital abnormalities of the uterus (e.g. uterus bicornis)
    h   Local injury

Clinical features:
  1 History: Careful history required:
    a   Nature of the bleeding: type, amount, duration of flow
    b   Date of last menstrual period
    c   Associated symptoms: bleeding tendency or associated pain or discomfort
    d   Associated medication
    e   General health

  2 Physical examination: Full physical examination, particularly looking for:
    a   Oedema of extremities
    b   Abdominal fullness: solid masses or ascites
    c   Abdominal tenderness
    d   Skin lesions that may be associated with vulval pathology
    e   Swelling or tenderness or discharge from Skene's or Bartholin's glands
    f   Pelvic examinations: for tenderness, induration, nodulation, swelling in pelvis

Investigation:
  1 Full blood count – check for anaemia

  2 Blood clotting tests

  3 Vaginal smears for cytological and microbiological examination

  4 Cervical smear cytology

  5 Diagnostic curettage under anaesthesia preferably performed between bleeding episodes

6 Radiological studies:
  a Plain abdominal film to reveal abdominal masses and fluid levels, if any
  b Hysterosalpingography – to visualize presence of polyps

Management:
  1 During an acute bleeding episode
    a Bed-rest and sedation
    b Place patient in Trendelenburg position and provide intravenous fluids and blood replacement

  2 Control the bleeding:
    a Surgical dilatation and curettage – this is both therapeutic and diagnostic
    b Control bleeding with oral progestagens – these suppress gonadotrophin production and thereby reduce endometrial vascular dilatation and permeability

  3 Specific management of dysfunctional uterine bleeding – see below

DYSFUNCTIONAL UTERINE BLEEDING

Definition: Abnormal uterine bleeding occurring during the reproductive years, that is not associated with an organic condition. It is a diagnosis of exclusion.

Pathophysiology:
Forms of dysfunctional uterine bleeding are:
  1 Anovulatory cycles (metropathia haemorrhagica) – about 75% of all dysfunctional uterine bleeding
    a Occurrence:
      i During the 1–3 years following puberty
      ii Occasionally during the reproductive years – in particular after a miscarriage or a pregnancy
      iii During the 3–5 years prior to the menopause (accounts for over half the total presentations)
    b Features:
      i Often initiated when a Graafian follicle has failed to rupture

    ii There is thus a prolonged period of oestrogen secretion and this enhances growth of the endometrium

    iii The blood oestrogen rises to a single peak, and then falls

    iv There is then a shedding of a thickened endometrium which has not undergone a secretory change as a Graafian follicle (producing progesterone) had failed to form

2 Ovulatory cycles:
Features: abnormal (dysfunctional) uterine bleeding is due to some abnormalities of ovarian function such as:

    a Variations in oestrogen/progesterone ratios

    b Defects in follicular maturation with an inadequate luteal phase, or

    c Persistence of an active corpus luteum

3 Pseudo-ovulatory cycles

    a Occurrence: Luteinization of the follicle is achieved but the follicle fails to rupture, owing to an inadequate mid-cycle peak in LH secretion

    b Features: Progesterone is secreted – hence most of the clinical features of ovulation do occur

Possible pathological changes in the endometrium:

1 Cystic glandular hyperplasia (as in 1 above)

    a Occurrence:

        i Excessive oestrogen secretion

        ii Absence of progesterone

    b Features:

        i Thickened endometriun

        ii Proliferation of glandular cells with cystic dilatation of the ducts

2 Follicular endometrium

    a Occurrence:

        i Oestrogen secretion not excessive, but

        ii Deficient progesterone secretion

    b Features: Endometrial histology resembles that of the follicular phase of the normal menstrual cycle

Clinical features:
1 History – often shows well-defined onset of symptoms:
    a If failure of ovulation: period may be delayed by up to a week. Vaginal bleeding then prolonged in duration
    b If ovulatory cycles: period begins at normal time but is heavier than usual

2 Examination: General and pelvic examination usually normal

Investigations:
1 Cervical smear cytology

2 Examination under anaesthetic (EUA) and dilatation and curettage (D and C) premenstrually with histological examination of the curetted endometrium. It is important to rule out organic causes – in particular malignancies of the genital tract

3 Hysterosalpingogram to demonstrate intrauterine abnormalities – present as filling defects

4 Haemoglobin concentration

Management:
1 General management
    a Oral iron supplements
    b Some degree of counselling – dysfunctional uterine bleeding often associated with a degree of emotional stress

2 Management of a bleeding episode
    a Replacement of blood loss may be necessary
    b If severe or prolonged bleeding give 0.05 mg ethinyl oestradiol every 6 hours. Once bleeding is controlled add 4 mg norethisterone per day to prevent excessive endometrial proliferation before withdrawal of oestradiol

3 Medical treatment: Combination of oestrogen and progesterone
    a Parous patient complaining of excessive or irregular bleeding

    i   0.05 mg ethinyl oestradiol for 17 days from first menstrual day, followed by 0.08 mg ethinyl oestradiol and 0.5 mg norethisterone for 9 days. Withdrawal bleeding occurs 2 days later

    ii Failing this treatment, 0.05 mg ethinyl oestradiol with 0.5 mg norethisterone from day 5 of the cycle for 21 days

  b Nulliparous women: Smaller dose of progestagen owing to risk of 'post pill infertility'. 0.05 mg ethinyl oestradiol for 19 days, and then 0.4 mg norethisterone added for 7 days. Withdrawal bleeding will occur 2 days later

4 Surgical treatment:

  a Curettage – many women's symptoms improve after curettage alone

  b Hysterectomy indicated only if medical treatment has failed, is contraindicated, or in the elderly patient

THE MENOPAUSE

The definition of menopause is the cessation of menstruation.

Physiological menopause: This is the gradual decline and final cessation of menstruation. It occurs between 44 and 52 years, but the time of onset and termination of the manifestation of the menopause varies. About 75% of women have the menopause between 44 and 50, the mean age being 48–49.

Artificial menopause: This may occur at any age, after menarche. It results from pituitary or ovarian destruction or ablation or serious pelvic disease.

Climacteric: This is the period surrounding the end of the reproductive phase of life and the word climacteric is often used to describe the 'change of life'.

Physiological changes (Figure 3.2):

1 Gonadal failure results in the 'breakdown' of the normal negative feedback relationship between the ovarian production of sex hormones, and of FSH and LH. This results in

  a Oestrogen depletion

  b Increased secretion of FSH and LH

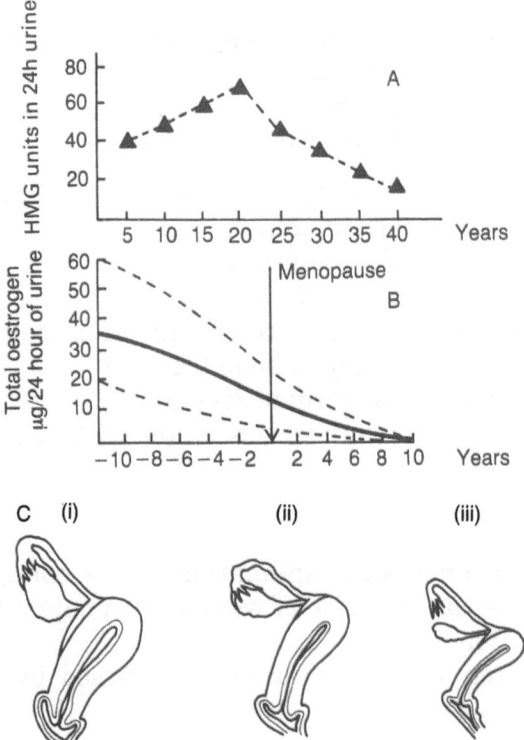

Figure 3.2   A *Gonadotrophin excretion during the climacteric and old age*. B *Excretion of total oestrogen in the urine during climacteric and old age*. C *Comparison of structure of female reproduction organs in (i) a nulliparous woman (20–30 years); (ii) a nulliparous 55 year old; (iii) an aged 75 year old.*

2   The severity of menopausal symptoms does not directly relate to hormonal changes, but women who are emotionally stable are less likely to have difficulty adjusting to the changes involved

Clinical features:
1   The two major features are vasomotor instability (manifested as hot flushes) and the eventual cessation of periods:
   a   Alteration and final cessation of the menstrual flow
   b   Vasomotor flushes and flushes, often accompanied by tachycardia, perspiration, and feelings of emotional tension. 'Hot flushes' may last for up to 15 minutes and occur several times a day

    c Vulvar and vaginal atrophy – may be accompanied by irritation and dyspareunia

    d Less specific symptoms may include emotional instability, digestive disorders and headache

2 Features that develop after the menopause: These are due to the deficiency of oestrogens

    a Atrophic (senile) vaginitis

    b Atrophic (senile) cystitis

    c Osteoporosis – may result in back pain, and an increased tendency to bone fracture

    d Arteriosclerosis; cardiac disease

    e Skin and hair changes associated with ageing

Management:

1 General

    a Reassurance and explanation of the changes taking place – the majority of women soon adjust to the new situation

    b Sedation if patient unduly distressed by vasomotor symptoms

2 Specific treatment

    a Treatment with oestrogens if it is necessary to reduce vaginal irritation, or dyspareunia

    b Analgesics where necessary for osteoporotic symptoms

Those women on long-term therapy should not have oestrogens alone. Unopposed oestrogen therapy causes hyperplasia of the endometrium and may be related to an increased risk of endometrial cancer.

Prognosis: Well-informed, emotionally stable patients usually adjust with minimal difficulty. Less well-adjusted women may undergo considerable distress over the menopause.

Causes of postmenopausal bleeding

Postmenopausal bleeding is abnormal flow from the genital tract 6 months or more following cessation of menses. This is a serious symptom and always must be fully investigated.

1 Iatrogenic – commonest:
  a Oestrogen replacement therapy. No demonstrable lesion

2 Vulval:
  a Carcinoma
  b Urethral caruncle
  c Urethral carcinoma

3 Vagina:
  a Senile vaginitis
  b Foreign body
  c Carcinoma

4 Cervix:
  a Polyp
  b Carcinoma

5 Body of uterus:
  a Carcinoma (up to 75% of cases arise in the postmeno-pausal years)
  b Sarcoma
  c Polyps
  d Submucous fibroid
  e Senile endometritis

6 Ovary:
  a Granulosa cell tumour
  b Thecal cell tumour

Diagnosis:
  1 General examination

  2 Examination of vagina and cervix – including a vaginal swab and a cervical smear

  3 D and C

Treatment:
  1 If malignancy found – definitive treatment required

  2 Atrophic vaginitis – dinoestrol cream

  3 Endometrial hyperplasia – hysterectomy

4

Gynaecological Infections; Sexually Transmissible Diseases

GYNAECOLOGICAL INFECTIONS

Presentation

1 Discharge
   a Urethral discharge may result from:
      i Gonorrhoea
      ii Non-specific urethritis
      iii Trichomoniasis
      iv Candidiasis
      v Local urethral lesions, e.g. herpes
   b Discharge from the vagina may result from:
      i Candidiasis or trichomoniasis
      ii Retained foreign body
      iii Allergic sensitization, e.g. to contraceptive creams, etc.
   c Discharge originating from the cervix may be either:
      i Due to infection – e.g. gonorrhoea, herpes
      ii Non-infective – e.g. cervical erosion or IUD

The discharge in all cases should be examined by:
   a Microscope and Gram-staining:
      i Discharge offensive – suggests trichomoniasis
      ii Discharge frankly purulent – suggests gonor-rhoea (Gram-negative intracellular diplococci)

    b Culture techniques by direct inoculation on to Stuart's transport media

2 Urinary symptoms: The commonest symptoms are dysuria and frequency

3 Swellings, often in the groin – inguinal lymphadenopathy

4 Other skin lesions (rashes, warts, sores), e.g. the primary lesion of syphilis

SEXUALLY TRANSMISSIBLE DISEASE

Sexually transmitted diseases (STD) are now among the commonest group of communicable diseases, and their incidence has been rising over the last twenty years. This increase in incidence has caused much concern and it is no exaggeration to say that in several parts of the world gonorrhoea is out of control. Similarly the world incidence of syphilis has also been rising. STD are a broad group of conditions caused by a wide variety of organisms:

1 Viruses:
- **a** Herpes simplex
- **b** Non-specific genital infections
- **c** Warts
- **d** Lymphogranuloma venereum
- **e** Molluscum contagiosum
- **f** Hepatitis B (some cases)

2 Bacteria:
- **a** *Neisseria gonorrhoeae*
- **b** Chancroid
- **c** Granuloma inguinale

3 Protists: Trichomoniasis

4 Spirochaetes: Syphilis

5 Fungi: Candidiasis

6 Arthopods:
- **a** Scabies
- **b** *Pediculosis pubis*

Incidence:
1 There are around 400 000 new presenters, cases that present to special clinics for sexually transmitted disease in England and Wales, and the number is steadily increasing

2 However the incidence of particular diseases may have declined, e.g. tertiary syphilis

3 The age group in which STD occurs most frequently is 15–50 years – the age of maximum sexual activity

Factors contributing to incidence:
1 Changes in sexual attitudes and behaviour

2 'Carriers' of infection themselves not displaying symptoms

3 Increased mobility of people

4 Increased insensitivity to antibiotics, e.g. β-lactamase production by some strains of *Neisseria gonorrhoeae*

5 Inadequate sex education

6 There is thought to be an increase in STD among contraceptive users particularly with oral contraceptives and the IUD

Clinical manifestations of sexually transmissible disease

Clinically, sexually transmissible disease may present locally as one or more of the following:
1 Genital ulceration

2 Vaginal discharge

3 Non-ulcerative genital lesions

4 Swellings (lymphadenopathy) in the groin

Causes of genital ulceration

1 Primary or secondary syphilis

2 Herpes simplex

3   Mechanical or chemical trauma

4   Secondary to scabies and pediculosis

5   Secondary to dermatoses, pyogenic infection etc.

6   Lymphogranuloma venereum

7   Chancroid

8   Granuloma inguinale

9   Behçet's disease

10   Allergy

11   Neoplasia

Causes of vaginal discharge

1   Physiological discharge (leukorrhoea)

2   *Trichomonas vaginalis*

3   *Candida albicans* vaginitis

4   Gonorrhoea

5   Non-specific urethritis

6   Herpes genitalis

7   Trauma, foreign body

8   Infected cervical erosions or polyps

9   Senile vaginitis

Causes of non-ulcerative genital lesion

1   Genital warts

2   Molluscum contagiosum

3   Balanoposthitis

4   Keratoderma Blennorrhagia (of Reiter's disease)

5   Fungal infections

6 Lichen planus

7 Psoriasis

8 Lichen sclerosis

9 Lichen simplex

NB Not all of these diseases are sexually transmitted

VAGINA

Normal defences:
1 Normal vaginal secretion is clear to white and is acid (pH 3–4.5) It contains Gram-positive commensals (*Lactobacillus acidophilus*) or Döderleins bacillus which convert glycogen into lactic acid

2 Cervical secretions are alkaline, but do not greatly affect the overall pH of the vagina

3 Before puberty the above defence mechanisms are not in operation

Pathogens:

In children, before puberty:
1 Foreign body

2 *Streptococcus aureus* and *Staphylococcus pneumoniae*

3 *Neisseria gonorrhoea*

4 Occasionally: thread worms

Causes of a non-infective vaginitis include:
1 Secondary to a cervicitis

2 Presence of foreign bodies

3 Chemical vaginitis, as in the use of irritants for douching

4 'Senile vaginitis' may occur after the menopause or post-oophorectomy

Clinical features:

1 Purulent discharge

2 Painful vagina

3 Dyspareunia

4 Dysuria

5 Frequency of micturition

Investigation of a vaginal discharge:
1 Full history, in particular enquiring for:
    a Onset and character of the discharge
    b Relation to coitus
    c Use of contraceptives or douches
    d Recent medications – especially vaginal applications
    e Recent pregnancy

2 Examination:
    a General examination and urine analysis
    b Pelvic examination:
        i Inspection and swabs, where indicated, of vulva, urethra and Bartholin's glands
        ii Speculum examination and swabs taken from cervix and posterior fornix (high vaginal swab) and cervical canal
        iii Examination of cervix for erosions, polyps or carcinoma
    c Note: Material from swabs should be:
        i Put in Stuart's transport media for culture for gonococcus (see below)
        ii Examined directly under the microscope for *Trichomonas vaginalis* and *Candida albicans*
A definitive diagnosis should form the basis of treatment.

Trichomonas vaginalis (Figure 4.1)

Microbiology: The causative organism is a flagellate protozoan. It is identified microscopically by a dark-ground illuminated wet preparation and is recognized by a terminal membranous stylus and four flagellae anteriorly. It can be cultured in a medium

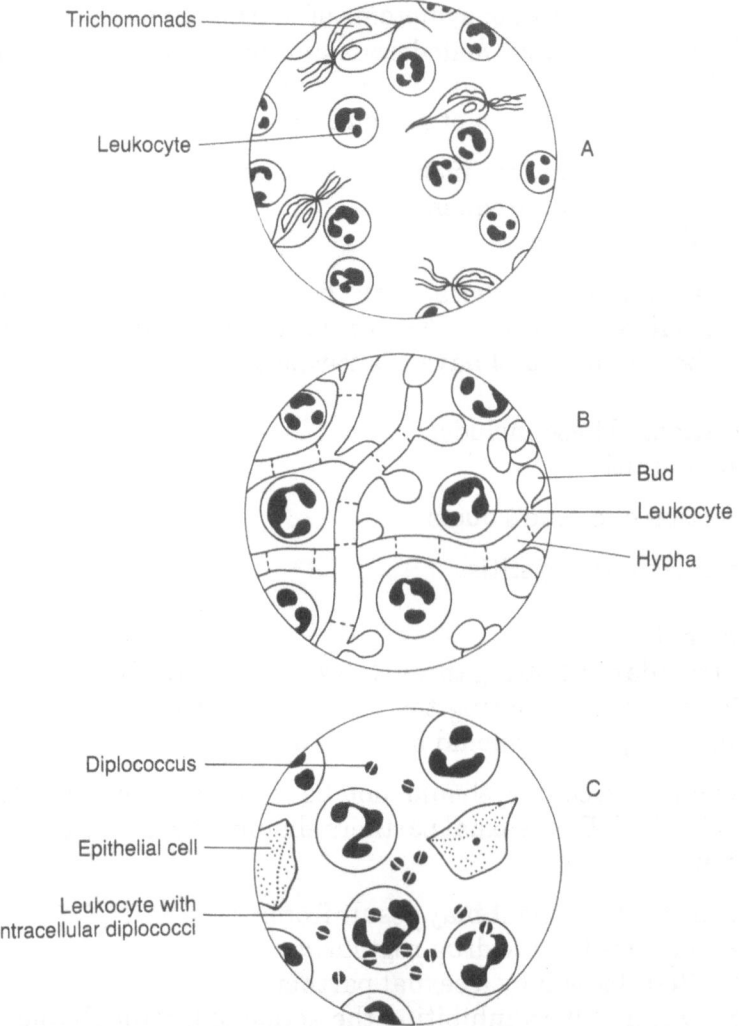

Figure 4.1   A *Trichomonas in vaginal secretions. Presence of numerous leukocytes implies inflammation; B* Candida albicans; *C Gram-negative intracellular diplococci:* Neisseria gonorrhoeae.

which contains proteolysed liver (Fineberg–Whittington medium). It is also found in the vagina of asymptomatic patients

Transmission: Most commonly, by sexual contact – men are frequently symptomless carriers. An appreciable number of women with trichomonal vaginitis will also have gonorrhoea.

Clinical features: Onset of symptoms often sudden and acute:
1 Discharge – frothy, malodorous and purulent usually yellowish green in colour. There may be secondary vulval oedema and irritation of surrounding skin

2 Additional symptoms include:
   a Dysuria and frequency
   b Dyspareunia

3 On examination, there may be additionally enlarged inguinal lymph nodes. The cervix and vaginal walls may also be inflamed and have a 'strawberry' appearance

Complications. These include:
1 Bartholinitis

2 Infection of Skene's ducts

3 Urethritis and cystitis

Management:
1 Metronidazole 200 mg three times a day, orally for 7 days, is effective in eliminating the parasite in over 80% of cases. The male partner should be treated with the same dose

2 Sexual intercourse should not be resumed until cure is established. Both sexual partners should attend for examination

3 Failure of treatment may result from:
   a Failure to take drug regularly
   b Re-infection by a sexual partner
   c Vaginal flora inhibiting the action of metronidazole

Monilial vaginitis (Thrush)

Microbiology: Causative organism is a Gram-positive fungus, *Candida albicans*, which forms long filaments that also produce clusters of spores. *Candida albicans* occurs as a frequent commensal in the genital tracts of both sexes. There is an increasing incidence of genital tract infection with fungi. This may relate to the increasing use of contraceptive pills and broad-spectrum antibiotics.

Transmission:
1 Infection may occur from the existing commensal flora (particularly from the gastrointestinal tract) of the patient, in particular:
  a If the patient is on contraceptive pill
  b If the patient has recently been treated with a broad spectrum antibiotic
  c During pregnancy
  d In patients with diabetes mellitus
  e In patients taking immunosuppressant drugs
  f In the presence of a foreign body in the vagina

2 Infection may also occur from a sexual partner or rarely via towels or toilet seats etc.

Pathology: The fungus usually infects skin and mucous membrane. However, systemic infection is known to occur, and may result in septicaemia and meningitis

Clinical features:
1 History:
  a Vaginal discharge – thick, whitish and cheesy
  b Vaginal and vulval itch and irritation

2 Examination:
  a Vulval skin may be tender, red, scaly and fissured
  b Vaginal wall may be covered with white plaques covering small haemorrhagic spots

Investigation:
1 Microscopy: Smears are Gram-stained and examined under a microscope. This should reveal mycelia and spores of the fungus

2 Culture on Sabauraud's medium

Management:
Satisfactory treatment depends on a definitive diagnosis
1 Basic treatment:
  a Two vaginal pessaries of 100 000 iu of nystatin inserted high in the vagina before bedtime for two weeks

    b Nystatin cream for other affected areas

2 Repeated relapses treated with a prolonged course of nystatin over several weeks

3 Patients in which candida is present in the intestinal flora or in the mouth, are treated with 500 000 units of oral nystatin three times a day for ten days

4 Other antifungal agents such as econazole and miconazole can be substituted for nystatin

Other forms of vaginitis

1 Chemical vaginitis – due to unsuitable chemicals used for douching or contraception

2 Foreign bodies, e.g. pessaries, contraceptive devices and tampons. Managed by removing the foreign body

3 Senile vaginitis – this results from atrophy of the vaginal epithelium. It may occur:
    a After the menopause
    b After oophorectomy
    c After an artificial menopause (see Chapter 3)

Clinical features:
1 Watery, often blood-stained discharge

2 Tenderness of vagina and dyspareunia

3 On examination – atrophic vagina and vulva with submucous haemorrhages

Investigation:
1 Swab to determine any infective organism

2 If discharge is blood stained, manage as for post-menopausal bleeding

Management:
1 Treat any infection that may be present

2 Treat atrophy of the vaginal epithelium:

a Lactic acid pessaries

b Local or low dose systemic oestrogens

CERVIX

Cervicitis (inflammation of the cervix) may either be acute or chronic. Acute cervicitis is considered later (see below).

'Chronic' cervicitis

Causes:

1 Inadequately managed acute cervicitis

2 Cervical erosion and cervical ectropion

Cervical erosion

In cervical erosion the stratified squamous epithelium which normally covers the vaginal portion of the cervix is replaced by columnar epithelium that is continuous with that of the cervical canal. It may be either congenital, or acquired, e.g. after parturition. It is one of the commonest disorders of the genital tract.

Clinical features:

1 Many patients are asymptomatic

2 Vaginal discharge – mucopurulent if there is concomitant infection

3 There may be slight postcoital bleeding

4 Urinary symptoms – frequency and dysuria

Differential diagnosis:

1 Cervical polyps

2 Carcinoma of the cervix

3 Cervical ectropion

Management: Two methods are in use – thermal cauterization or cryosurgery of the lesion. NB: Carcinoma of the cervix must first be excluded.

Cervical ectropion

This occurs when the cervix is badly damaged during delivery or surgery. The normal endocervical columnar epithelium is exposed at the os. A small ectropion does not usually cause symptoms but a large one may produce a mucoid discharge. It is healed by excising the lips of the tears and the cervical canal is restored by suturing the raw surfaces.

GONORRHOEA

This is an acute infectious disease that involves the urogenital tracts of both sexes. Severe forms of the disease can invade the blood stream and spread to joints, meninges etc. It has a high (and increasing incidence) of around 70 000 new cases per year in England and Wales. Factors that have increased the incidence of gonorrhoea include:

1   Altering patterns of sexual behaviour

2   It is often asymptomatic in the female and she can unknowingly continue to transmit the disease

3   Increased use of the contraceptive pill instead of mechanical or barrier methods of contraception; the latter it is thought provides a degree of protection against sexually transmitted infections

4   An increasing number of β-lactamase-producing, penicillin-resistant strains

Microbiology: *Neisseria gonorrhoeae*: a Gram-negative bacillus. Is a strict parasite and is soon destroyed outside the body.

Transmission:
1   Sexual intercourse – commonest and most important mode of transmission

2   By transvaginal infection of the conjunctiva of a neonate during passage through the birth canal, causing 'Ophthalmia Neonatorum'

Clinical features:
History:
1 Short incubation period of 2–8 days

2 50% or more of infected women may be asymptomatic and are only detected by contact-tracing

3 The remainder may have only mild, and relatively non-specific symptoms, e.g.
   a Vaginal discharge
   b Burning micturition and/or frequency
   c Back or pelvic pain

Examination: A careful examination, especially of the pelvic organs, is indicated:
1 Inguinal nodes: may be enlarged and inflamed

2 Labia may be reddened. Vaginal discharge may be present

3 Urethra may be tender and there may be a periurethral abscess

4 Bartholin's gland – abscess formation

5 Cervix may be inflamed on speculum examination

6 Tenderness may occur due to oviduct or ovarian involvement on bimanual pelvic examination

Investigation: Diagnosis is dependent upon demonstrating that the organism is present in stained slides and on culture. It cannot be made on clinical examination alone and cannot be excluded merely by use of a high vaginal swab.
1 Collection of bacteriological samples. A sterile platinum loop should be used to obtain samples from:
   a Urethra
   b Cervix
   c Vagina
   d Rectum – over 50% of infected females have rectal gonorrhoea

2 Microscopy: This should reveal under high-power Gram-negative, intracellular diplococci

3   Culture: This is necessary for the following reasons:
  a   False positives occur where Gram-positive organisms fail to stain, or where related Gram-negative organisms are present
  b   Rectal and oropharyngeal specimens are more reliably cultured than seen under microscopy
  c   A diagnosis based on stained smears alone may be inadequate for legal purposes

Culture requires a blood-enriched agar, e.g. heated blood agar, brought up to incubator temperature just before inoculation. Where inoculation cannot be performed immediately, the organism must be kept alive using Stuart's transport media, and collected using a carbon-containing swab stick. Incubation is in the presence of 10% carbon dioxide for 48 hours.

4   Identification:
  a   Appearance – gonococcal colonies are small, circular and transparent. The cocci are Gram-negative and stain black with a special stain
  b   Biochemical tests are needed to confirm diagnosis. When incubated with glucose, maltose and sucrose *Neisseria gonorrhoeae* ferments only glucose
  c   Immunological identification employs immunofluorescent antibody, but the validity of the particular test used has recently been questioned

5   Tests for sensitivity to antibiotics, in particular for β-lactamase-producing strains employs disc techniques

Serological testing for gonorrhoea: Gonococcal complement fixation test on blood serum attempts to measure the presence of antibody in the patient's serum. However, this test is unreliable and of no use in diagnosis of early acute infections. It is not, therefore, a substitute for thorough microbiological investigation

Management:
1   Antibiotics:
  a   Single intramuscular injection of 4.8 megaunits procaine penicillin, preceded by 1 g probenecid orally

b Other regimes are indicated where:
  i There is a history of penicillin sensitivity
  ii Suspicion of early syphilis
  iii Evidence for infection with β-lactamase produc-
    ing strains of *Neisseria gonorrhoeae*
  iv Treatment failure with penicillin

These regimes include:
  i Spectinomycin hydrochloride – single intramus-
    cular injection of 2 g
  ii Kanamycin sulphate – single intramuscular
    injection of 2 g

2 Management of complications – see below

3 General patient care: Patient should be told of the infectious nature of the disease. Sexual intercourse should be
prohibited until cure has been established. Instructions
about simple hygiene should be given.

4 Contact tracing: The patient should be questioned in confi-
dence about the source of the infection, and about anyone to
whom the patient may have subsequently passed on the
infection

5 Follow-up:
  a Re-examination about 1 week after treatment for
    repeat smears and cultures. These may be repeated
    after 2 weeks
  b After 3 months: repeat tests, and perform serological
    tests to exclude syphilis before discharge from clinic

Complications of gonorrhoea:
1 Local complications:
  a Skenitis
  b Bartholinitis
  c Proctitis

2 Pelvic complications (see below):
  a Salpingitis
  b Abscesses, in particular involving ovaries and/or ovi-
    ducts

    c Peritonitis – may be pelvic or generalized

3 Metastatic complications (gonococcaemia)
    a Gonococcal septicaemia: fever, generalized malaise, flitting arthritis, red papular skin rashes
    b Gonococcal arthritis
    c Other complications include meningitis, pericarditis, endocarditis and liver (perihepatitis) involvement. These may have fatal consequences

SALPINGITIS

Pathology: This is an inflammation of the oviduct, via one of the following routes:
1 Ascending infection from vagina, and/or uterus e.g. gonorrhoea

2 Spread from pelvic peritoneum

3 Haematogenous spread

Presentation may be acute or chronic.

Acute salpingitis

This may follow:
1 Infection by *Neisseria gonorrhoea*

2 After childbirth or termination of pregnancy – usually streptococci, staphylococci and *Escherichia coli* are causative organisms

3 As a complication to pelvic operation

Clinical course: Is as shown in Table 4.1 (page 71)

Clinical features of acute salpingitis:
History:
1 Severe constant lower abdominal pain

2 Purulent vaginal discharge

3 Dysuria and frequency of micturition

Table 4.1   Clinical course of acute salpingitis

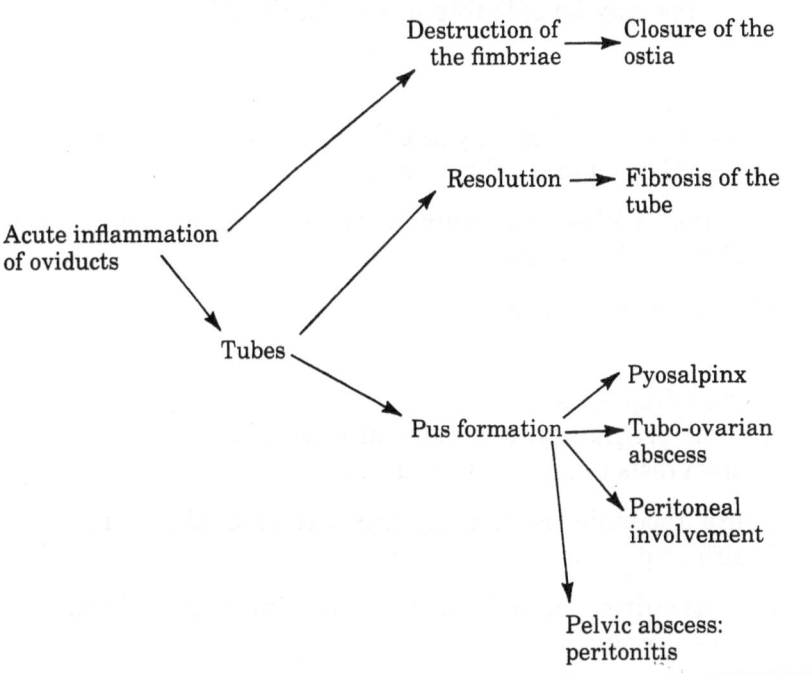

4 Chills and fever are common

5 Menses may be increased in amount and duration

Examination:
1 General: Marked pyrexia and tachycardia

2 Abdomen:
   a Lower abdomen acutely tender in both quadrants
   b Signs of peritoneal irritation (rebound tenderness and guarding)

3 Pelvic examination:
   a On speculum examination of the cervix a purulent discharge may be seen
   b Tender and painful cervix
   c Uterus may be fixed and/or retroverted

> **d** Adnexae very tender bilaterally and adnexal swellings may be palpable in the pouch of Douglas

Investigations:
1 Causative organism may be cultured from pus or from cervical canal or from urethral swabs

2 Elevated white cell count with a marked polymorphonuclear leukocytosis

3 ESR is usually elevated

Differential diagnosis:
1 Ectopic pregnancy. Stigmata of infection, viz. pyrexia and leukocytosis usually less marked

2 Acute appendicitis. Usually the pattern of abdominal pain is different

3 Diverticulitis – may be difficult to differentiate from left-sided salpingitis

4 Acute pyelonephritis. Pelvic signs are usually not marked

5 Ovarian tumour: tension, rupture, haemorrhage. Swelling is usually on one side only

Management: Patient should be admitted to hospital
1 Conservative management is indicated in the acute phase:
   a Symptomatic measures include:
      i Bed-rest
      ii Local heat and ample analgesia
   b Treatment with broad spectrum antibiotics should be started once samples for bacteriological examination have been taken:
      i Ampicillin, 500 mg six-hourly for *at least* two weeks
      ii Metronidazole 200 mg eight hourly is often added to this regime
      iii Adequate antibiotic management is important to minimize the chances of chronic infection

2 Surgical management is indicated in the event of failure of medical treatment, or if the diagnosis is in doubt at laparotomy cultures are taken and any pus is drained

Complications:
1 Acute:
  a Pelvic abscess: this may form a fluctuant and tender swelling
  b Pyosalpinx: Either may rupture and cause a pelvic or general peritonitis

2 Chronic infection which may result in sterility, especially from inadequate antibiotic treatment

3 Recurrent attacks are frequently due to re-infection

Chronic salpingitis

Pathology: May or may not follow from a history of acute salpingitis. Broad spectrum of pathology, including:
1 Residual mild chronic inflammation of the oviducts – this may result in fibrosis and thickening of the oviducts with formation of adhesions and lead to infundibular thickening that damages and blocks the tube

2 Hydrosalpinx: formation of clear exudate within a blocked segment of tube caused by an infection of low virulence

3 Tubo-ovarian abscesses; pyosalpinx. Accumulation of pus in an occluded and infected segment of oviduct
Severity of symptoms probably depends on virulence of the organism, the immune state of the patient and history of previous medication.

Clinical features of chronic salpingitis:
History: Varied:
1 In the mildest cases, the patient may be asymptomatic and present with sterility due to occluded tubes

2 A large number of cases present with
  a Chronic lower abdominal pain
  b Menstrual disturbances including menorrhagia and secondary dysmenorrhoea

    c Deep dyspareunia
    d Persistent mucopurulent vaginal discharge

3 In severe cases the patient may be debilitated and additionally complain of loss of weight

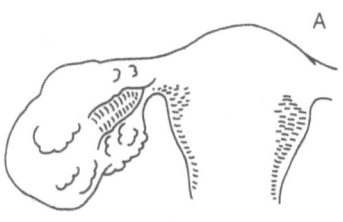

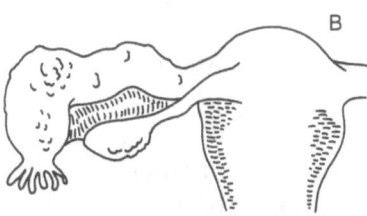

Figure 4.2   *Chronic inflammation of the adnexae. A Bilateral pyosalpinges – occlusion of fimbriae, distension of tubes, perisalpingitis and adhesions to the ovary; B Tuberculous salpingo-oophoritis. Thickening of the tube, narrowed ostium, perisalpingitis and peri-oophoritis.*

Examination: Uterus may be fixed and retroverted with tender adnexae in which swellings may be present (Figure 4.2)

Investigation: Laparoscopy may be needed to confirm the diagnosis. The principal differential diagnosis is endometriosis (see Chapter 5).

Management: Dependent on severity of symptoms:
  1 Trial of conservative management may be made if:
     a No definite mass is detected
     b Symptoms are mild
  This consists of a full course of antibiotic therapy. Surgical

attempts to restore patency of the oviducts may be attempted, but results are usually poor

2 Surgical management – bilateral salpingectomy and total hysterectomy, attempting to conserve at least one ovary

Pelvic cellulitis

Pathology: Infection of the connective tissue surrounding pelvic organs. Often secondary to:
1 Salpingitis

2 Cervical lacerations following parturition or criminal abortion
Pathological pattern may be acute or chronic

Clinical features: Vary in severity and time course. Symptoms and signs classically occur between 7 and 10 days after infection

History: Malaise, pyrexia, and dull-aching pelvic pain

Examination: The uterus may be fixed, retroverted, or displaced to one side, and an indurated mass may be felt in the adnexae

Management:
1 Symptomatic – analgesia for relief of pain

2 Antibiotics – Ampicillin 500 mg 6 hourly. If clinical condition does not improve within 2 days, or if infection is due to anaerobes, drug may be changed. Metronidazole 200 mg 8 hourly is often used in conjunction with ampicillin

3 If an acute abscess develops, drainage may be necessary through a posterior colpotomy

SYPHILIS

Syphilis is a chronic infectious disease caused by the spirochaete *Treponema pallidum*, which is capable of involving practically every organ system. The disease is characterized by florid manifestations in some stages, and long periods of asymptomatic latency in others. The disease is relatively uncommon; about

2000 new cases are reported each year.

Microbiology: The causative organism is the spirochaete *Treponema pallidum*. This is a spiral organism, demonstrated by microscopic observation under dark-ground illumination.

Transmission: This is almost always as a result of sexual contact; the organisms gain entry either through minor breaks in the epithelium, or by penetrating intact mucosa. The treponeme is a delicate organism and cannot survive exposure to vaginal secretions or drying and becomes established only if brought into close contact with relatively neutral sites. Patients are maximally infectious in the primary and secondary phases (see below) of the disease, but sexual transmission and transmission of treponemes to the fetus in pregnant women becomes increasingly less likely in the latent phase of the disease.

Pathology: There are three recognizable stages:
1 'Primary' stage: this is characterized by lesions appearing at the site of infection

2 'Secondary' stage: the lesions are generalized but most affect the skin and mucous membranes

3 'Tertiary' stage: gummatous and destructive lesions occur, after an asymptomatic period of many years

Primary syphilis

A 'primary' change develops at the site of pathogen entry, usually on the labia minora or cervix in the female. It may have a (variable) incubation period of up to three months. It begins as a dull red macule which erodes to give a painless solitary indurated ulcer about 1.0 to 1.5 cm which heals spontaneously within 2–12 weeks

Secondary syphilis

Dissemination of the spirochaetes occurs rapidly after the initial inoculation, but overt symptoms may take eight weeks or more to appear:

1 General manifestations:
  a Mild 'flu-like illness: pyrexia, sore throat, headache
  b Generalized lymphadenitis

2 Skin lesions occur in over 70% of cases. They vary, but usually are discrete, painless, reddish and are not vesicular. The lesions (usually macular or papular) are up to 2 cm in diameter, and are distributed symmetrically, especially over the abdomen, arms and palms

3 Lymphadenitis may occur in around 40% of cases. The occipital, epitrochlear and cervical nodes may become enlarged, rubbery and discrete

4 Less common manifestations include:
  a Uveitis
  b Hepatitis
  c Central nervous system involvement

NB. Both primary and secondary syphilis disappear, even if untreated, between 3 and 12 months, whereupon the disease enters a latent stage.

'Latent' syphilis

This follows remission of the clinical features of secondary syphilis. It is characterized by:
1 Positive serological tests for syphilis

2 Absence of clinical evidence of the disease on physical examination

3 Negative cerebrospinal fluid tests

4 Normal cardiac and aortic shadows on radiographic screening

It is important to be able to diagnose latent syphilis as its treatment will prevent onset of manifestation of the 'tertiary' lesions

Tertiary syphilis

This may or may not follow the latent phase. It is characterized by the gumma. Features of gumma are: they are solitary or grouped; they are large and destructive lesions, and tend to erode at the edges and heal at the centre.

1   Cutaneous manifestations:
  a   Skin: gummata may occur as groups of painless, asymmetrical, indurated lesions localized to defined areas of the body
  b   Subcutaneous tissue: these begin as smooth swellings below the skin and break down to produce a gummatous ulcer
  c   Mucous membranes: gummatous lesions may occur in the mouth, the floor of the nose, or the tongue

2   Bones: Lesions especially in the tibia, skull, or femur, may be symptomless, or cause deep pain and radiological changes. They may involve either:
  a   The periosteum – gummatous periosteitis causes proliferation of bone tissue beneath the periosteum. This may be palpable, and is responsible for the 'sabre tibia'
  b   The bone itself: gummatous osteitis in contrast, causes destructive lesions, often seen as 'punched out' lesions in the skull

3   Liver:
  a   Multiple gummata may occur in the liver, especially in the early stages and produce hepatomegaly
  b   During healing, fibrous tissue is formed, and this contracts, resulting in destruction of the architecture of the liver, which divides into large irregular lobes (*'hepar lobatum'*)

4   Cardiovascular system: The treponeme damages the walls of the aorta and aortic valves, resulting in:
  a   Aortic incompetence: this produces the classical cardiovascular symptoms and signs, namely, breathlessness, ankle oedema, retrosternal pain, and features of a 'water-hammer' pulse
  b   Coronary obstruction: damage to the aortic wall may cause obstruction of the mouths of the coronary artery resulting in angina pectoris
  c   Aortic aneurysms: these result from damage to the

elastic wall of the aorta. They may be asymptomatic and only be detected on radiological examination, or may present as dyspnoea and chest pain, or present catastrophically if they rupture

5 Nervous system: Treponemes gain access to the central nervous system at an early stage of infection, but overt clinical symptoms of central nervous system involvement are rare before the tertiary stage. Pathological changes can be classified into:

    a Meningovascular syphilis – this results from damage to the meninges and blood vessels serving the central nervous system. It may be asymptomatic and detected only on physical examination. Clinically evident symptoms may result from:

        i Cerebral lesions, resulting in headache, diplopia, mental changes, 'upper' motor neurone paresis

        ii Spinal lesions, resulting in radicular pain across the chest, and progressive paresis of the legs

    b Parenchymatous neurosyphilis ('general paralysis of the insane': GPI) results from involvement of nerve cells themselves within the central nervous system. There is progressive mental and physical deterioration. This culminates in complete dementia and paralysis with double incontinence

    c Tabes dorsalis may occur by itself or with GPI. It results from destruction of the dorsal columns of the spinal cord. It presents with 'lightning pains' in the legs, numbness and anaesthesia in the lower limbs and a characteristic gait

'Congenital' syphilis

This refers to syphilis transmitted from the mother to the fetus via the placenta. Clinical features can be divided into early manifestations, reflecting acute infection, and late manifestations

1 Early congenital syphilis:

The infected infant:

    a May be born with a red, dusky rash and covered with

bulbous lesions ('syphilitic pemphigus')
or

b Lesions that resemble those of secondary syphilis may appear a few weeks after birth

2 Late manifestations may occur beyond the first year of life, with gummata formation, and consequences similar to those of tertiary syphilis

3 Long-term stigmata: These are the consequence of damage to developing tissue *in utero* by the treponemes and include:

a Thickening ('bossing') of the bones of the skull
b Notching of the teeth ('Hutchinson's teeth')
c VIIIth nerve deafness

Investigation

Treatment should almost always be based on objective diagnosis. Antibiotic therapy based on clinical grounds alone is rarely justified, and attempts should always be made to obtain a definitive diagnosis.

1 Dark-ground microscopy: This is the principal means of diagnosis in the primary and secondary stages. Exudate is obtained from the lesion by cleaning the lesion with normal saline and then scarifying the lesion with a wire loop. If satisfactory amounts cannot be obtained by this method, aspirated fluid from an infected lymph gland may be used instead. The exudate is examined under high power, employing an oil immersion and a dark field

2 Serological tests: These are used to diagnose latent or tertiary syphilis, or when diagnosis based on dark-ground microscopy is in doubt. The classical Wasserman reaction (WR) and Kahn tests are being progressively replaced by the following more specific tests for treponemal antibody:

a Venereal Disease Research Laboratory (VDRL) test: Heat inactivated serum from the patient is mixed with a freshly prepared suspension of cardiolipin–lecithin–cholesterol antigen on a glass slide. Flocculation is determined microscopically. The VDRL test is versatile and in common use

b Treponemal immobilization (TPI) test: Virulent organisms are mixed with heat-inactivated serum in the
presence of complement. If the serum is positive, i.e.
contains immobilizing antibody, the treponemes will
be seen to be immobilized under dark-ground microscopy. This test is available only in relatively few
centres and is rarely used now.

c The fluorescent treponemal antibody adsorbed (FTA-
ABS) test. *Treponema pallidum* are fixed to a microscope slide. Patient's serum is added to the slide: if
antibodies are present, they attach to the treponemes.
This can be detected by examination under fluor-
escence microscope. Often this test is the first to
become positive in early syphilis

d *Treponema pallidum* haemagglutination (TPHA) test:
Tanned sheep erythrocytes are sensitized by treponeme breakdown products. When mixed with positive
serum from a patient, agglutination of these cells
occurs. This test is extremely sensitive and may
remain positive even after the conclusion of successful
treatment

Management of syphilis

Penicillin remains the drug of choice for all stages of the disease.
There is no objective evidence for any significant development of
penicillin resistance by the treponeme – the organism does not
appear to have the genetic capacity to develop resistance.

Regimes: Unlike for the treatment of *Neisseria gonorrhoeae*, successful therapy relies upon the maintenance of a sustained low
level of circulating antibiotic:

1 A single intramuscular injection of procaine penicillin G
600 000 units per day for 10–15 days will maintain a therapeutic blood level for a sufficient period
or

2 An intramuscular injection of benzathine penicillin G (2.4
megaunits) will maintain an effective serum level for up to
2 weeks and so will minimize the number of injections
required

When used in the therapy of syphilis, penicillin may provoke a Jarisch–Herxheimer reaction. In the early stages symptoms are usually mild, with rigors, headache, joint pain and local swellings. However they may be more severe when they occur in latent cases and can be prevented by previous treatment with a steroid (e.g. prednisone 5 mg four times a day beginning the day before penicillin is given). For patients who are hypersensitive to penicillin:

1 Tetracycline hydrochloride 500 mg four times a day by mouth for 14 days
   or

2 Erythromycin 500 mg four times a day by mouth for 14 days.

   In patients who have had syphilis for one year or more the above regimes may be given over a period of 3–4 weeks.

SEXUALLY TRANSMISSIBLE VIRUS INFECTION

The following viruses may be transmitted by sexual contact:

1 Herpes simplex virus

2 Human wart virus

3 Hepatitis B virus

4 Epstein–Barr (EB) virus

5 Molluscum contagiosum

6 Cytomegalovirus

Herpes simplex virus

Pathology: Causative agent is Herpes virus type II. It is demonstrable by electron microscopy and can be grown on tissue culture. Besides causing systemic illness and vesicular lesions in the genital area, the virus has a predilection for nerve roots and may remain quiescent at this site for long periods. Herpes virus type I is usually isolated from lesions around the mouth and eye and transmitted by direct contact or by droplet infection.

Clinical features: Transmission occurs by close contact with a person who has a recurrent lesion. Primary attack is usually the most severe and includes:

1 Severe constitutional upset

2 Local pain and tenderness in the genital area and local lymphadenitis

3 Vesicles or ulcers in the genital areas

4 With widespread vulval involvement dysuria is common

5 In pregnant women, the virus may cross the placental barrier and cause severe and often fatal illness in the fetus. The usual source of infection of the neonate is the mother's infected genital tract at delivery

Investigation: Inoculation of swabs taken from the herpetic lesions onto tissue culture of human embryo (lung) cells. The virus is characterized by antigenic typing, pock size, cytopathic effect and the appearance of infected cells in tissue culture

Management:

1 Symptomatic; washing lesions with normal saline daily; severe cases may require hospital admission

2 Management of intercurrent bacterial infection with antibiotics

3 Note: there is no objective evidence that application of local or systemic nucleosides such as idoxuridine hastens recovery or prevents recurrence

4 There is good evidence that a young woman with a history of repeated attacks of genital herpes has a larger chance of developing cervical carcinoma. It is therefore advisable to perform yearly cervical smear cytology in such cases

Human wart virus (genital warts)

Pathology: Genital warts are caused by the human papilloma virus. The virus is transmitted both by the sexual route, and by skin contact. The incubation period is on average about 3 months.

Clinical features: Lesions ('condyloma acuminatum') are red-brownish raised areas often occurring in crops in the genital region, especially on vagina, cervix and anus. In men the warts appear most often on the glans penis and on the prepuce

Investigation: Biopsy may be needed to exclude a more sinister lesion

Management:

1 Scrupulous cleanliness of the affected areas, with local application of podophyllin in alcohol

2 More resistant warts may require ablation with electro-cautery or cryosurgery

Hepatitis B virus

Sexual transmission: Hepatitis B antigen ('Australia' antigen) is demonstrable in seminal fluid and in vaginal secretions as well as in serum of infected patients. Sexual transmission is also suggested by a high incidence of 'Australia' antigen in attenders at venereal disease clinics, especially among homosexual patients.

Clinical features:

1 The majority of contacts may be completely asymptomatic, but be infective for long periods

2 The remainder will show the clinical features of acute infective hepatitis

Investigation: All patients suspected of, or who have had Hepatitis 'B' should be tested for HB_e antigen (HB_eAg). This is a diagnostic indicator, and also tests for infectivity. Normally levels have disappeared between 3 to 6 months after recovery, but a small number of cases remain infective indefinitely.

Infectious mononucleosis

The Epstein–Barr (EB) virus may be spread by sexual contact as well as by the respiratory route. In most cases it causes pyrexia, lethargy and a sore throat with generalized lymphadenopathy.

The Paul–Bunnell test is positive. Management is expectant, and complete recovery occurs.

Molluscum contagiosum

The virus of molluscum contagiosum is spread by close body contact often under moist conditions. It causes small circular, pink and painless lesions on the vulva and pubic area. These hyperplastic and hypertrophic lesions are treated with a phenol stick. Electrocautery can also be used. Also exclude any other coexistent sexually transmissible disease.

Cytomegalovirus (CMV)

CMV is endemic in all human societies, but the disease is usually asymptomatic and can be transmitted as a primary infection during pregnancy or postnatally. CMV, like infectious mononucleosis, may be a 'kissing disease'. The consequences of congenital CMV can be numerous and include hepatosplenomegaly, microcephaly, purpura, chorioretinitis and uveitis. It is a much commoner cause of congenital malformation than rubella.

TROPICAL SEXUALLY TRANSMISSIBLE DISEASE

Chancroid

Microbiology: Causative agent is a Gram-negative coccobacillus, *Haemophilus ducreyi*. Incubation period is within a week to ten days. It occurs in tropical countries particularly in areas where living standards are low

Clinical features:
1 Painful shallow ulcers on the genitalia which bleed easily on contact occur

2 The inguinal lymph nodes may be enlarged and painful

Investigation:
1 Causative organisms may be difficult to demonstrate by microscopy

2 Culture may be possible using defibrinated blood

Management:
1 Frequent application of saline dressings to the lesions
2 Sulphadimidine 5 g per day (in divided doses) for two weeks
3 Aspiration of inguinal abscesses should these form

Lymphogranuloma venereum (LGV)

Microbiology: Causative organism is *Chlamydia trachomatis*, which is an obligate intercellular parasite. LGV is mainly acquired during sexual intercourse with an affected partner

Clinical features:
1 Small painless ulcer which rapidly disappears, but is followed by:

2 Enlargement of the inguinal lymph glands with possible abscess development (bubo formation) and lymph vessel obstruction. Inguinal lymphadenitis is more common in men than women

Investigation:
1 Microscopy of smear of material from abscesses

2 Culture has to be on a cellular medium (McCoy cells)

3 Test for antibody to chlamydia group of organisms

Management:
1 Tetracycline 500 mg 6-hourly for 14 days or sulphadimidine 5 g per day (in divided doses) for 16 days

2 Repeated aspiration of buboes as necessary

Granuloma inguinale (Donovanosis)

Microbiology: Causative organism is a Gram-negative intracellular bacillus *Donovania granulomatis*. Transmission is thought to occur by both sexual and non-sexual means.

Clinical features: Painless, gradual development of a slowly spreading granulomatous ulcer found in the ano-genital region. It spreads along the skin folds to involve the whole inguinal region. The regional lymph nodes are not enlarged, painful or tender.

Investigation: Organisms may be demonstrated as intracellular bodies on Giemsa staining. Culture is difficult. It is important to perform serological tests since granuloma inguinale is associated with chancroid, syphilis and LGV in about 20% of cases.

Management: Tetracycline 2 g per day for 2 weeks, or ampicillin 500 mg orally six hourly for 2 weeks.

NON-SPECIFIC GENITAL INFECTION: 'NON-SPECIFIC URETHRITIS'

Microbiology: *Chlamydia trachomatis* is the causative organism. Mycoplasmas and *Ureaplasma urealyticum* may be implicated in a few cases.

Clinical features: Often asymptomatic in the female, and presentation is then through contact tracing. Where the disease is symptomatic, the commonest presenting symptoms are:

1 Vaginal discharge, due to cervicitis

2 Frequency and dysuria due to cystitis or urethritis

3 Pelvic inflammatory disease may result from ascending infection

4 Later complications include Reiter's disease

Investigation: The causative organism is cultured from a cervical swab immersed in Hank's transport medium.

Management: Tetracycline 1 gram per day (divided doses) for 14 days. Patient advised to avoid alcohol and coitus during this period.

TUBERCULOUS INFECTION OF THE GENITAL TRACT

Pathology: Results from haematogenous spread of *Mycobacterium tuberculosis* from an infected focus elsewhere. Oviducts are involved in most cases resulting in a chronic infection that produces thickening and fibrosis of the walls of the tubes. More acute infection may result in a tuberculous pyosalpinx. Endometrial involvement may result in tubercles in the compact layer, and eventually destruction of the endometrium.

Clinical features: Symptoms vary with the severity of the infection. They may range from:

1 Acute form – fever, night sweats, malaise, loss of weight – relatively uncommon presentation

2 Mild cases – history of sterility or irregular uterine bleeding. Pelvic examination may be negative

3 More severe cases – amenorrhoea may result from endometrial destruction
Pelvic pain with adherent pelvic masses may be present

Investigation: Endometrial curettings should be subjected to:

1 Histological examination using Zeihl–Neilsen staining

2 Culture on Loeffler's medium with drug sensitivity tests
In case of negative results, it is still possible that infection is present in the oviducts which may require diagnosis by a hysterosalpingogram. This may reveal:
1 Tubal strictures
2 Deformation of the ampulla
3 Calcification in the oviduct walls and in the ovary
The original focus of infection should be found

Management:
1 Medical management is tried first
   a Rifampicin 500 mg per day
   b Isoniazid 300 mg per day
   c Ethambutol 400 mg per day for the first month
The result of chemotherapy should be assessed by endometrial histology and culture, and follow up should continue for 5 years after supposed remission

2 Surgical management: This is indicated only in the failure of medical treatment or if there are caseous abscesses or fistulae. It should be accompanied by chemotherapy

ACQUIRED IMMUNE DEFICIENCY SYNDROME (AIDS)

Definition

No unequivocal marker to identify individuals with AIDS. Working definition adopted by Centers for Disease Control, USA requires:

1 Biopsy proven Kaposi's sarcoma and/or life-threatening infection verified by biopsy or culture

2 Patient under 60 years old with no history of underlying immunosuppressive disease or treatment

These criteria only identify the most severe and later manifestations of the disease.

Epidemiology and mortality

Over 5890 cases of AIDS have now been notified in the USA by the Centers for Disease Control, Atlanta (September 1984). Of these cases, 2690 (46%) have died. However, at least 70% of AIDS cases are dead within the first year of diagnosis.

In the UK there have been a total of 81 cases of AIDS of which 36 have died. (Figures collated by the Centre for Disease Surveillance, London, 30 September 1984.)

Sporadic cases of AIDS have been reported as early as 1978. The growth rate of the epidemic has been exponential with some indication that the rise is now levelling off.

The majority of AIDS patients are aged between 20 and 49 years.

Distribution

1 USA: Marked geographical clustering of the epidemic within the USA. Approximately 65% of AIDS cases reside in the metropolitan areas of:
   a New York
   b San Francisco
   c Los Angeles

These conurbations have large homosexual communities.

**Reported cases of AIDS by area of residence,
June 1981–April 1983 USA**

	Cases	*% Total cases*	*Cases per million population*
New York City	603	46	66
San Francisco	164	13	51
Los Angeles	93	7	12
Miami	50	4	31
Newark	32	2	16
Elsewhere USA	358	28	2
Total	1300	100%	6

2 Outside USA: AIDS has been identified in 33 countries, notably in France, Germany, Belgium and Haiti. The number of cases in December 1983 totalled 446.

The 81 cases reported in the UK have been categorized as follows by the Centre for Disease Surveillance:

Risk group	*Male*	*Female*	*Total cases*	*Total deaths*
Homosexuals/bisexual	70*	0	70	30
Haemophiliacs	3	0	3	1
Cases associated with Central Africa	0	3†	3	3
Other and unknown	3	2	5	2
Total	76	5	81	36

*One of these was also an IV drug abuser
†One of these was a white British woman who had visited Central Africa

Risk Groups

Population	*% of first 2000 cases in USA*
1 Homosexual or bisexual men	71
2 IV drug addicts	17
3 Haitian	5
4 Haemophiliacs	0.7

Four other 'at risk' groups have now been identified:

5 Sexual partners of AIDS patients

6 Transfusion patients

7 Infants of females with AIDS

8 Cases associated with Central Africa

Transmission of the 'AIDS agent' appears to be by:

1 Sexual contact

2 Parenterally (Factor VIII concentrate, other blood products, contaminated needles)

3 Perinatally

Aetiology

Causative agent still unknown.
Strongest candidate is Human T-cell leukaemia virus (HTLV III)/Lymphadenopathy associated virus (LAV)
Other possibilities which have been considered include:

1 Cytomegalovirus (CMV)

2 Epstein–Barr virus (EBV)

3 Hepatitis-B virus

4 Retrovirus

Other factors such as genetic predisposition are also involved but exact relationship is still not established.

Clinical features

Opportunistic infection common in AIDS:

1 *Pneumocystis carinii* pneumonia – the most frequent cause of death in AIDS

2 Cytomegalovirus infection

3 Candidiasis

4 Mycobacter

5 Herpes simplex

6   Cryptococcal infection

7   Toxoplasmosis

Malignant neoplasms common in AIDS:

1   Kaposi's sarcoma: Most common malignant tumour in AIDS patients. Occurs in 25% of cases. Kaposi's sarcoma is a very rare tumour in the USA (<1% of all cancers). In Africa it is 200 times more frequent. The relevance of African Kaposi's sarcoma to AIDS has received much speculation recently. It has been hypothesized that AIDS originated in rural equatorial Africa where lack of medical facilities and the presence of many endemic infections would allow an immune deficiency syndrome to go unrecognized

2   Other tumours reported in a *few* AIDS patients which may also be manifestations of the syndrome include non-Hodgkin's lymphoma, Burkitt's lymphoma, intra-cranial lymphoma and squamous cell carcinoma

Signs and symptoms of AIDS:

1   Dyspnoea with or without pulmonary infiltrates – most commonly due to *Pneumocystis carinii*

2   Fever – common causes include pneumocystis, mycobacterial, cryptococcal and CMV disease

3   Neurological dysfunction – primarily due to cryptococcal meningitis or toxoplasmosis infection

4   Diarrhoea – bacterial or parasitic infection

5   Significant weight loss

6   Skin lesions – Kaposi's sarcoma

7   Chorioretinitis – can be caused by CMV and toxoplasma

Persistent generalized lymphadenopathy and AIDS: Less than 10% of patients with persistent generalized lymphadenopathy develop AIDS

Investigations and diagnosis

No reliable diagnostic blood test or set of tests for AIDS. Diagnosis established on the basis of clinical events. Laboratory investigations:

1 Blood picture:
   a Leukopenia (WBC count <1000–4500 cells/mm^3)
   b Lymphopenia (1500 cells/mm^3)
   c Mild–severe anaemia (Hg <12g/dl)

2 Bone marrow biopsy: accumulations of histiocytes or granulomas – can indicate fungal or mycobacterial infection

3 Serological evaluations: over 90% of AIDS patients have antibodies against Hepatitis A or B, CMV, EBV

4 Microbiological studies:
   a CMV in urine or throat cultures.
   b Over 30% of AIDS patients have CMV in blood culture
   c Over 80% have EBV in throat culture

5 Radiographic studies: abdominal CT, ultrasound or lymphangiography may reveal enlargement of abdominal lymph nodes and spleen

6 Immunological evaluation:
   a Decreased T helper/T suppressor (largely T helper depletion)
   b Decreased T lymphocytes
   c Decreased natural killer cell numbers and function

Management:

1 Malignant complications: e.g. Kaposi's sarcoma:
   a Radiation for local disease
   b Chemotherapy with VP16, vinblastine
   c Immunotherapy with interferons

2 Infective complications: treated conventionally, e.g. *Pneumocystis pneumonia*:
   a High dose co-trimoxazole
   b Pentamidine

3 Underlying disease: No specific treatment possible until the cause of AIDS has been found. Attempts at 'immune reconstitution' have been unsuccessful:
  a Interferons
  b Interleukin 2
  c Thymic hormone
  d Bone marrow transplantation

Directions and goals

1 Care for patients with the disease

2 Rational public health measures for avoiding exposure in high risk groups

3 Research into the causative agent:
  a potential for developing effective treatment
  b potential for developing vaccine

5

Benign and Malignant Neoplasms of the Female Genital Tract

The female genital tract is the site of a large number of tumours of considerable diversity.

1   Benign – the commonest are:
   a   Fibromyomata of uterus and ovaries
   b   Physiological ovarian cysts
   c   Pseudomucinous and serous cystadenomas of the ovary
   d   Adenomyosis of the uterus
   e   Endometriosis
   f   Endometrial and endocervical mucous polyps
   g   Bartholin's cysts and abscesses

2   Malignant – the commonest are:
   a   Carcinoma of the ovary
   b   Carcinoma of the endometrium
   c   Carcinoma of the cervix

FIBROMYOMATA

Incidence: Fibromyomata are the commonest of all uterine tumours.

Aetiology:

1   Cause is unknown

2   Arise from smooth muscle cells (myomata) during reproduc-

tive life. Large tumours also contain fibrous tissue (fibromyomata)

3 Suggested that fibroids are the result of an abnormal response to oestrogen

4 Very common in negresses

Symptoms:
1 Can be asymptomatic and may not require treatment

2 Menorrhagia is the commonest symptom caused by enlargement of the endometrial surface, endometrial hyperplasia or by interference with uterine contractions

3 Pain is unusual with uncomplicated tumours

4 Pressure symptoms – commonly on bladder causing retention of urine. Large tumours may cause varicosities and leg oedema

Signs:
1 Large tumours are felt on abdominal examination as firm, rounded, smooth swellings which rise up out of the pelvis

2 Pelvic examination – cervix moves laterally with the tumour mass

Pathology:
1 Fibroids are usually multiple and may grow to enormous sizes

2 Seedlings first appear in the uterine wall and compress the surrounding tissue to form a capsule

3 The tumours are gradually excluded from their intramural position towards the uterine cavity to become subendometrial, or towards the peritoneal surface to become subperitoneal

4 Both submucous and subserous fibroids may become pedunculated

5 Degenerative changes may occur:
   a Atrophy

   b Hyaline degeneration

   c Cystic degeneration

   d Calcareous degeneration

   e Red degeneration (necrobiosis)

   f Malignant degeneration – sarcomatous change occurs in less than 0.5%

6 Torsion of a pedunculated tumour may occur acutely but rarely

7 Infection may occur after torsion

8 Impaction of a fibroid in the pelvis may cause urinary retention

Diagnosis:

1 Ultrasound

2 X-ray – plain film of the abdomen may show calcification

3 Laparotomy

Treatment:

1 Many fibroids are asymptomatic and do not require treatment

2 If menorrhagia is the predominant symptom and fibroids are small diagnostic curettage may suffice

3 Surgery:

   a Polypoid tumours may be removed by vaginal myomectomy

   b Abdominal myomectomy is indicated in nulliparae under 40 or in parous women who want further children

   c Abdominal hysterectomy is the definitive treatment

ENDOMETRIOSIS

The occurrence of endometrial tissue in sites other than the uterine cavity.

Aetiology:

1 Usually manifests itself between the ages of 30 and 40 years

2   Between 50 and 70% of the women are nulliparous

3   It is mostly confined to Caucasian women and is more prevalent in high socio-economic groups

4   Theories of the histogenesis of endometriosis:
   a   Retrograde menstruation and implantation of viable cells on the ovaries, peritoneum of the pouch of Douglas etc.
   b   Metaplasia – both epithelial and stromal cells of the endometrium have a common precursor in the coelomic epithelium and adjacent mesenchyme. Endometriosis could result from abnormal differentiation of primitive cells
   c   Mechanical transplantation into scars at the time of surgery, e.g. hysterectomy or hysterotomy. It very rarely follows Caesarean section.
   d   Venous or lymphatic spread. This could account for the rare pulmonary endometriosis

Pathology:
1   The ectopic endometrium menstruates causing a sterile inflammatory reaction, and dense adhesions

2   The commonest sites are both ovaries which may show merely surface deposits or contain 'chocolate' cysts (containing old menstrual blood) of various sizes

3   The next commonest sites are the pouch of Douglas and uterosacral ligaments

4   The small and large intestine, the appendix, umbilicus and rectum may be involved. Rectal bleeding or painful defaecation may occur at the time of the menses. It may be difficult to differentiate clinically from carcinoma

5   Fibroids and endometriosis occur together in about 40% of cases

Symptoms and signs:
1   Pain:
   a   Dysmenorrhoea
   b   Mittelschmerz or ovulation pain

    c Dyspareunia

2 Infertility

3 Heavy often irregular menses

4 Symptoms of intestinal obstruction may occur

Diagnosis:
1 The presence of tender nodules on the uterosacral ligaments

2 Fixed retroversion of the uterus and bilateral cystic ovaries

3 Laparoscopy

4 For histological diagnosis both glandular and stromal tissue must be present

Treatment:
1 Medical treatment:
    a Continuous progestagen therapy (e.g. didrogesterone 20–30 mg daily) for six to nine months
    b Danazol 200–800 mg daily (depending on severity) for four to six months. Danazol interferes with gonadotrophin secretion, inhibits ovulation, and abolishes menses

2 Surgical treatment:
    a Conservative surgery: small deposits are diathermied or excised. Large chocolate cysts require excision
    b Hysterectomy and bilateral salpingo-oophorectomy

ADENOMYOSIS

A condition where islets of endometrial tissue, glands and stroma are found deep in the uterine wall. Three types are recognized:
1 Internal endometriosis

2 External endometriosis

3 Stromatous endometriosis – a rare malignant condition of the uterus

Features:

1 It tends to occur in older, multiparous women than external endometriosis

2 Clinical features include:
   a Severe menorrhagia
   b Secondary dysmenorrhoea
   c Gradually enlarging tender uterus

3 Symptomatically it is difficult to differentiate from other causes of uterine enlargement with or without menorrhagia

4 The uterus may be diffusely thickened or there may be localized swellings resembling leiomyomas

Treatment:

1 Medical treatment – hormones only after a firm diagnosis has been made

2 Surgical treatment – hysterectomy with, where possible, conservation of ovaries

SWELLINGS AROUND THE VULVA

1 Congenital – hypertrophy of clitoris or labia minora

2 Trauma – leading to a haematoma

3 Vascular – varicose veins sometimes present during pregnancy

4 Infections – pruritus vulvae

5 Cysts – sebaceous, inclusion dermoid, Wolffian duct remnants, endometriosis

6 Condylomata:
   a Condylomata acuminata (viral warts)
   b Condylomata lata (secondary syphilis)
   c Molluscum contagiosum

7 Primary chancre (primary syphilis)

8 Enlargment of Bartholin's gland:

   a Bartholin's adenitis – abscess formation may occur – caused by gonococci, staphylococci, streptococci or Gram-negative organisms

   b Tumours – adenomas or adenocarcinomas

9 Vaginal and uterine causes – prolapse, polypus, inversion of the uterus

10 Urethral and paraurethral conditions – prolapse, diverticulum or caruncle, cyst of Skene's paraurethral glands

11 Inguinal causes – inguinal hernia, or hydrocoele of canal of Nuck

12 Benign neoplasm – papilloma, fibroma

13 Malignant neoplasms – squamous cell carcinoma, melanoma, sarcoma

PHYSIOLOGICAL OVARIAN CYSTS

Follicular cysts ('cystic ovary')

1 Usually single and small, and less than 5 cm in diameter except when caused by overuse of fertility drugs (e.g. clomiphene or gonadotrophins)

2 Most cysts arise from atretic follicles

3 They do not become malignant and in the majority of cases have no clinical significance

4 The cyst wall is composed of an incomplete layer of granulosa cells enclosed in the theca layers of the follicle

5 Management – most cysts require no treatment

Lutein cysts

1 The ripe corpus luteum occasionally produces a cyst usually less than 5 cm in diameter

2 Short periods of amenorrhoea may be present

3 Occasionally haemorrhage occurs into a cyst and may cause bleeding into the peritoneal cavity. Thus, mimicking an ectopic pregnancy

Theca-lutein cysts

1   Both the granulosa and theca interna cells lining a follicular cyst may become luteinized

2   Theca-lutein cysts are found in hydatidiform mole and choriocarcinoma and in some patients taking fertility drugs drugs

3   Both ovaries are usually enlarged up to 10 cm in diameter

CARCINOMA OF THE VULVA

Incidence: Very rare: Primary cancer of the vulva accounts for about 5% of all genital cancer. Most gynaecologists will see about two cases each year. The majority of cases occur in women aged between 50 and 70 years.

Aetiology: Little is known although preinvasive conditions such as Bowen's disease and Paget's disease are found in some patients. It has been suggested that poor hygiene and chronic infection are predisposing factors.

Symptoms:
1   There is usually a long history of pruritus vulvae

2   Presence of a lump, which may be multiple

3   Pain, discharge of serous fluids and bleeding at the margin of the carcinomatous area

4   Urinary symptoms

5   Loss of weight

Signs:
1   Evidence of leukoplakia or other vulval dystrophy

2   Usually a typical epitheliomatous ulcer but sometimes a hypertrophic or 'cauliflower' lesion is seen

3   Often situated on the anterior half of the vulva usually affecting the labia majora. Lesions may occur in the labia majora (40% cases), posterior commissure (28%), clitoris (17%) or labia minor (15%)

4 Inguinal lymph nodes may be palpable

Pathology: Over 80% are well-differentiated squamous carcinomata. A preceding premalignant lesion occurs in around 40% of these cases. In the remainder, the malignant transformation appears without an apparent premalignant stage. Other rare forms include:

1 Basal cell carcinomas

2 Adenocarcinomas from Bartholin's gland or adjacent glands are rare (0.2% of all genital carcinomas)

3 Mixed tumours e.g. melanoblastomas

Spread: Metastasis is mainly by lymphatic channels. There are five main groups of lymph nodes arranged in two layers, a superficial and a deep layer that have communications between them. The superficial and deep nodes drain into the gland of Cloquet. Distant metastases are rare.

Clinical Staging (TNM classification):

T – Primary tumour

T_1 – Tumour confined to vulva, 2 cm or less in diameter

T_2 – Tumour confined to vulva, more than 2 cm in diameter

T_3 – Tumour of any size with extension to the lower urethra and/or vagina and/or perineum and/or anal orifice

T_4 – Tumour of any size with extension to upper urethra and/or mucosa of bladder and/or rectal mucosa or fixed to pelvic wall

N – Regional lymph nodes

N_0 – No nodes palpable

N_1 – Nodes palpable in either groin, not enlarged, mobile

N_2 – Nodes palpable in either one or both groins, enlarged, firm and mobile

N_3 – Fixed or ulcerated nodes

M – Distant metastasis

M_0 – No clinical metastasis

M_{1a} – Palpable deep inguinal lymph nodes

M_{1b} – Other distant metastasis

Management:

1 Biopsy of suspicious skin area

2 If carcinoma *in-situ* is found a simple vulvectomy is indicated

3 If invasive – radical vulvectomy
  a Excision of the vulva, including the labia
  b Inferior margin of the vaginal mucosa and perimeatal mucosa
  c Bilateral block dissection of the inguinal lymph nodes

4 Radiotherapy is disappointing – it may cause very painful necrosis and ulceration

Prognosis: The overall five year survival rate from radical vulvectomy is about 70%. Prognosis depends on the degree of lymph node invasion and on the radicality of treatment. When the deep nodes are involved the 5 year survival rate is about 40%.

CARCINOMA OF THE VAGINA

Incidence: Very rare and accounts for only 1–2% of all genital tract malignancies. It usually develops about 10 years after menopause. Secondary cancer of the vagina is commoner than primary.

Aetiology:

1 No common aetiological factors

2 There may be a history of dysplastic lesions of the vagina

3 An occasional case has been reported following the development of a traumatic ulcer created by a chronically retained ring pessary

4 A few cases, usually clear-cell adenocarcinomas, have been found in young women whose mothers received stilboestrol during pregnancy

Symptoms and signs:

1 Painless bleeding or watery discharge

2 Vaginal mass

3 Pruritus

4 Constipation

Most of the lesions occur in the upper half of the vagina usually on the posterior wall

Pathology: Histology is that of a squamous-cell carcinoma in the majority of cases. A few adenocarcinomas developing from the epithelium of the remnants of Gartner's duct may be found.

Spread: Lymphatic spread occurs late involving the external iliac and hypogastric nodes. If spread involves the vulva the inguinal lymph nodes may be involved. Ureteric involvement with obstructive renal failure is the commonest cause of death. Haematogenous spread is rare.

Clinical staging:
    Preinvasive carcinoma *in-situ*, intraepithelial carcinoma
    Invasive carcinoma of the vagina
    Stage I: Carcinoma limited to vaginal wall
    Stage II: Carcinoma has involved the subvaginal tissues but has not extended on to the pelvic wall
    Stage III: Carcinoma has extended on to the pelvic wall
    Stage IV: Carcinoma has extended beyond the true pelvis or has involved the mucosa of the bladder or rectum

Diagnosis:
  1 A positive cytological smear suggests the diagnosis if cervical cancer is excluded

  2 The vagina is then painted with Lugol's solution and the unstained areas subject to colposcopy and biopsy

  3 It may be difficult to identify the vagina as being the primary organ concerned if the lesion has spread into the cervix or the vulva

Management:
  1 Biopsy of ulceration or infiltration recognized on vaginal inspection. Frequent follow-up is required

2   Surgery – radical abdominal hysterectomy and pelvic lymphadenectomy including removal of vagina

3   Radiotherapy – intracavitary radium and external radiotherapy

4   Combination of surgery and radiotherapy

Prognosis: Depends on the type, location and extent of the tumour. With appropriate treatment the 5 year survival rate is about 35%.

Secondary cancer of the vagina

1   Local spread of cervical cancer

2   Metastases from ovarian, Fallopian tube cancer or choriocarcinoma – rare

3   Vaginal involvement occurs in around 10% of endometrial adenocarcinoma

CARCINOMA OF THE CERVIX

Incidence: The cervix is the most commonly affected by gynaecological cancer if carcinoma *in-situ* is included. It causes about 2200 deaths each year in England and Wales. About half of these occur between the ages of 45 and 65. Preinvasive carcinoma of the cervix occurs mainly before the menopause, on average about 13 years before invasive carcinoma.

Aetiology: Exact aetiology is unknown but several factors are involved:

1   Coitus at a relatively early age

2   Disease is very rare in virgins

3   Incidence rises with the number of pregnancies and particularly first pregnancy at an early age

4   Incidence is higher in the lower income socio-economic groups

5 Disease is very rare in Jewish communities and may also be rare in other communities that practise male circumcision but the evidence is unclear

6 Incidence is higher in promiscuous females

7 Carcinogens: Herpes simplex virus (type 2) has been implicated

Symptoms and signs:
Carcinoma *in-situ* and micro-invasive carcinoma have no clinical features – these conditions are detected only by screening methods. Invasive carcinoma may be characterized by:

1 Abnormal bleeding, postcoital, intermenstrual or postmenopausal

2 Foul discharge

3 Pain indicates widespread involvement and usually only occurs in the terminal phase of the disease

4 Bladder and rectal problems or fistulas are late symptoms

5 In late cases uraemia may develop due to obstruction of the ureters by paracervical spread

On clinical examination of a woman with invasive cancer the indurated lesion usually appears as rough and friable often with extensive local ulceration, or an exfoliative growth. Rectal examination may reveal parametrial spread and extension to the pelvic walls or lymph nodes

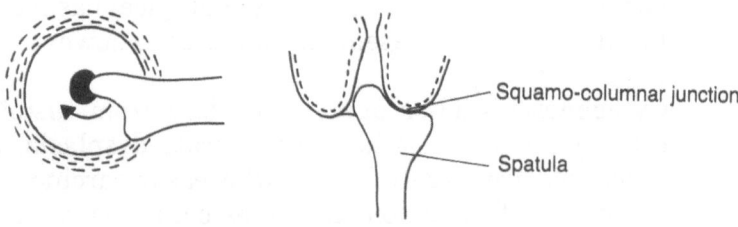

Figure 5.1   Obtaining a cervical smear

Screening:

1 Cervical smear test – this can detect carcinoma *in-situ* or
   Stage 0 carcinoma of the cervix before there is any visible
   evidence of the lesion. The technique for taking a cervical
   smear with an Ayres spatula is shown in Figure 5.1. In-
   terpretation of the stained smear is reported usually as one
   of five grades; it is believed that squamous carcinoma of the
   cervix passes through precursor stages characterized by
   progressive disturbances in maturation of the squamous
   epithelium:

   Grade I   Normal smears
   Grade II   Slight atypia (as seen in metaplasia or infec-
               tion)
   Grade III   Dysplasia. If mild, differentiation of the epi-
               thelium is preserved with only delayed matura-
               tion and increased mitosis of the basal cell
               layer
   Grade IV   Marked dysplasia with a few suspicious malig-
               nant cells with poorly differentiated nuclei
   Grade V   Very marked dysplasia with many malignant
               cells showing nuclear changes and basal dys-
               karyosis

   Mild dysplasia may be reversible. However, severe dysplasia and carcinoma *in-situ* are regarded as precancerous.
   Grades I and II are regarded as negative or benign, Grade
   III as suspicious and Grades IV and V as positive or malignant. At least 50% of severe dysplasias progress to invasive
   cancer.

   In carcinoma *in-situ*, a large number of cells with cytological abnormalities occur with a loss of differentiation into
   normal stratification. Mitoses and atypical basal cells occur
   in all layers. The basement membrane, however, remains
   intact.

   Cytogenetic studies agree with data from histology and
   cytology. Thus in mild and moderate dysplasia, the cells
   concerned are largely diploid, whereas in carcinoma *in-situ*,
   a number of aneuploid cell lines occur, with structurally
   abnormal karyotypes. The latent period from severe dysplasia to invasive cancer breaching the basement membrane varies from several months to ten years. Suspicious

and positive smears should be repeated and the patient referred for colposcopic examination of the cervix.

Cervical screening usually starts at ages 25–30. A second smear is performed one year later and thereafter smears are usually taken at 5 year intervals to the age of 70.

2 Colposcopy is an outpatient non-invasive procedure. The colposcope is a low-power binocular microscope that allows a magnification of ×6 to ×40 of the cervix. Acetic acid (3%) is applied to the cervix and causes the epithelium to swell and makes it easier to recognize. Certain basic appearances of the cervix can be recognized
   a Native squamous epithelium – stains brown with Schiller's iodine
   b Native columnar epithelium – seen as an erosion and has a 'grape-like' appearance on application of 3% acetic acid
   c Typical transformation zone – intervenes between the native squamous and native columnar epithelium
   d Atypical transformation zone – the area in which squamous cell carcinoma arises

3 Diagnostic cone biopsy is indicated if:
   a There is abnormal cytology but normal colposcopy
   b The upper limit of the lesion in the cervical canal cannot be visualized or is obscured by cervical stenosis
   c There are positive endocervical curettings

Pathology: Carcinoma of the cervix is usually of the squamous type and around 95% of cases begin in the region of the squamo-columnar junction. However, about 5% derive from columnar epithelium – adenocarcinomata (predominant type in very young women). The tumours may vary from well-differentiated squamous cell carcinomas to poorly differentiated anaplastic lesions.

Spread: Spread is usually by direct invasion of the paracervical tissues, parametrium, uterus and vagina. The ureters are involved in over 50% of patients. Rapid lymphatic spread through the rich lymphatic supply of parametrial tissue occurs by embolization and haematogenous spread to the liver, lungs and skeletal system is late and comparatively rare.

Clinical features:

1 Premalignant and early malignant lesions are asymptomatic and so screening is the only method of detection in these instances

2 A brownish discharge may be noticed when the tumour begins to ulcerate

3 Late symptoms reflect spread of tumour, and include:
   a Pain in the pelvis
   b Difficulty with micturition and defaecation
   c Oedema of the lower extremities

4 The most common cause of death is ureteric involvement resulting in obstructive renal failure

Staging: Clinical staging – Some centres still use the FIGO rather than the TNM classification
   Preinvasive carcinoma
   Stage 0   Intraepithelial carcinoma (carcinoma *in-situ*)
                Invasive carcinoma
   Stage I   Carcinoma strictly confined to the cervix (extension to the corpus should be disregarded)
   Stage Ia   Cancer cannot be diagnosed by clinical examination (early stromal invasion, occult cancer)
   Stage Ib   All other cases of Stage I
   Stage II   Carcinoma extends beyond the cervix but has not extended to the pelvic wall. Carcinoma involves the vagina but not the lower third
   Stage IIa   No obvious parametrial involvement
   Stage IIb   Obvious parametrial involvement
   Stage III   Carcinoma has extended on to pelvic wall. On rectal examination there is no cancer-free space between the tumour and pelvic wall. Tumour involves lower third of vagina. Presence of hydronephrosis or non-functioning kidney
   Stage IIIa   No extension on to the pelvic wall
   Stage IIIb   Extension on to the pelvic wall

Stage IV   Carcinoma has extended beyond the true pelvis or has involved the mucosa of the bladder or rectum

Management: Intraepithelial and micro-invasive carcinoma (Stage 0)

1 Local excision. A biopsy preferably minimized with the help of colposcopy

2 Conization of the cervix. The immediate risks of cone biopsy are bleeding and infection. Long-term risks include cervical incompetence, infertility, fibrosis leading to stenosis and dystocia in labour

3 Ablation by hot or cold cautery (including the CO_2 laser)

4 Hysterectomy – usually reserved for:
   a Those in whom another indication is present (e.g. fibroids)
   b Those who show abnormal smears and colposcopy after conization
   c Those who have lesions in the upper vagina

Investigation of patients with invasive cancer of the cervix

1 Full medical examination

2 Blood – full blood count, urea and electrolytes

3 Chest X-ray

4 Intravenous pyelography if indicated

5 Lymphangiography if indicated

6 Ultrasound scan

7 Examination under anaesthesia – to stage the extent of the growth and to decide on the method of treatment. Dilatation and curettage along with biopsy of obviously malignant cells from the cervix is also performed. Cystoscopy may also be undertaken

8 X-ray skeletal survey and isotope scans of bone and liver – may help to identify distant metastases particularly in cases of recurrence

Treatment of carcinoma of the cervix

Surgery: Hysterectomy with the removal of a wide vaginal cuff is usually reserved for selected operable cases. Radiotherapy is usually the treatment of choice. In advanced stages a combination of surgery and radiotherapy may be used. The choice between surgery and irradiation is often difficult because 5-year survivals do not greatly differ. Choice may depend as much on preference or availability. Side effects of irradiation are covered below. Surgical management may have a slightly higher mortality, and certainly a higher morbidity. However, it does provide prognostic information as it involves removal of the pelvic lymph nodes.

Radiotherapy:
1   Intracavitary radium supplemented by deep X-ray therapy
    a   Stockholm technique – high intensity radiation of short duration given 3 times in 3 weeks (repeated applications must be made)
    b   Paris technique – low intensity radiation given continuously for up to one week
    c   Manchester technique – predetermined doses delivered to fixed points in the pelvis, usually by two insertions with an interval of about a week between them. Recently caesium-137 a byproduct of nuclear energy programmes has tended to supplant radium

2   External irradiation – high-voltage machines – is used mainly for palliation in late stages of the disease

3   Interstitial irradiation using radium needles, cobalt or colloidal gold applied to paracervical tissues – not a popular technique

4   Transvaginal X-rays – superseded by external irradiation

The exact choice of which combination of therapy to employ depends to some extent upon the centre. Nevertheless the following trends apply for management of the different stages:
1   Stage I: Note that in these cases there remains a 15–20% risk of deep pelvic lymph node involvement that may

require radical surgical or irradiation treatment. However in Stage Ia lesions, the lymph node involvement is less than 2%, and some centres perform simple total abdominal hysterectomy to reduce the morbidity and mortality due to radical treatment.

2 Stages Ib, IIa and IIb: are treated by radical hysterectomy and/or primary irradiation

3 Stage III: Primary irradiation is indicated for lesions that have reached the pelvic wall or lower third of the vagina

4 Stage IV: Irradiation is usually ineffective and ultraradical exenteration procedures are performed if metastases remain confined within the pelvis

Details of the operative procedures are outlined in Chapter 9
Complications of radiotherapy:
1 Early side effects:
    a Nausea and vomiting
    b Weight loss
    c Dysuria and urinary frequency
    d Erythema of skin and subsequent pigmentation
    e Proctitis

2 Late side effects:
    a Tissue fibrosis
    b Haemorrhagic cystitis
    c Small or large bowel stenosis or fistulas
    d Vaginal stenosis
    e Leukopenia – seldom severe

Postoperative care and follow-up: Details may vary slightly with centre, but the following general points apply:
1 Patients in less fortunate socio-economic groups will need social support to assist their recovery and adjustment back to normal life

2 Full pelvic examination and cytological tests are required every 3 months for the first 3 years, and every 6 months thereafter. After 5 years, annual follow-up may be sufficient

3 Weight, haematocrit, urine analysis and other specific lab-

oratory tests as indicated should be performed at each visit

4   Symptoms reported that are referable to the urinary system
    or lower gastrointestinal tract should be investigated with
    cystoscopy and proctoscopy

Recurrences

A recurrence is said to occur if a lesion is demonstrated after a 6
month–5 year period during which the patient is free of disease.
After 5 years or more after treatment, they are termed 'late
recurrences'.

Clinical features: Most recurrences develop on the pelvic wall
and result from lymph node involvement. They are commonest
in the first two years. Diagnosis is difficult as there is usually
post-treatment scarring.

Management: Prognosis is poor, and management is difficult. If
the original management was surgical, irradiation may be tried.
Further irradiation after primary irradiation depends on the tolerance of the surrounding normal tissue, and in any case, recurrence after irradiation implies poor radiosensitivity of the
original lesion.

Terminal care

Considerable psychological and clinical skill is demanded with
careful, but adequate use of narcotic analgesia. Death is most
frequently from obstructive renal failure due to ureteric involvement.

CARCINOMA OF THE CERVICAL STUMP

Supracervical hysterectomy is an operation now not performed
but many women may have undergone this 20 or more years ago
and the incidence of cervical carcinoma from the stump is between 4 and 8%. Management is as for cervical carcinoma.

CARCINOMA OF THE CERVIX DURING PREGNANCY

Diagnostic procedures (including cone biopsy) are as for the non-pregnant patient:

1   Severe dysplasia; carcinoma *in-situ*. Pregnancy need not be interrupted and vaginal delivery is acceptable in the absence of obstetric complications. Definitive management is required about 6 weeks after delivery

2   More-advanced lesions, however, require prompt management. Pregnancy has to be terminated, although a delay of a few weeks may be permissible to allow the fetus to become capable of extrauterine survival in advanced pregnancy. Radical hysterectomy should then follow Caesarian section

Prognosis: Success in treatment over the last decade or so is the result of early cytological detection rather than through improvements in the management of advanced lesions. By convention, a cervical cancer is considered 'cured' if no metastases or recurrence have occurred by five years after initial treatment. Approximate 5-year survival rates from large series treated by radiotherapy are:

Stage I: 75% 5 year survival
Stage II: 55% 5 year survival
Stage III: 30% 5 year survival
Stage IV: 10% 5 year survival

The best results from surgery appear to be the same as the best results from radiotherapy.

HYPERPLASTIC AND PREMALIGNANT CHANGES IN THE ENDOMETRIUM

Cystic glandular hyperplasia

This results from continuous and unopposed oestrogen stimulation of the endometrium.
This results in:

1   Enlargement of the endometrial glands resulting in a 'swiss-cheese' hyperplasia

2   Proliferative-type endometrial stroma

This type of hyperplasia is not premalignant:

1   Glands and stroma are clearly separate

2   Endometrium and myometrium are distinct

However, cystic glandular hyperplasia can progress to adenomatous hyperplasia.

Atypical adenomatous hyperplasia

1   Glands are hypertrophied and hyperplastic and crowded together. Epithelial cells remain regular but may be stratified with frequent mitoses

2   Stroma is reduced

3   Areas of benign squamous metaplasia occur

The lesion must be regarded as premalignant as 6–12% of cases may develop into endometrial carcinoma over 1–10 years.

Endometrial dysplasia

This may be thought of as a stage intermediate between adenomatous hyperplasia, and endometrial adenocarcinoma:

1   Cells may be poorly differentiated with polymorphic nuclei and abnormal mitotic figures

2   There is proliferation of cells into the lumina of the endometrial glands

3   The stroma is reduced, but the basement membrane remains intact

Over 50% of the lesions of this kind progress to adenocarcinoma within 1–3 years. The histological change is irreversible.

Management: Adenomatous hyperplasia is responsive to progesterone. Also, it may be shed at menstruation. Prophylactic hysterectomy may be considered if the woman is within or beyond her reproductive years.

CARCINOMA OF THE ENDOMETRIUM

Incidence: Endometrial cancer is now more common than invasive cervical cancer in the UK. Patient age at onset is much higher than is the case with cervical cancer; the majority of patients with endometrial cancer are postmenopausal.

Aetiology: A number of factors have been implicated:
1 Age – 75% of patients with endometrial cancer are over 50. Many of these reached the menopause after the age of 50

2 Parity – incidence equally common in parous and nulliparous patients

3 Previous menstrual abnormalities – about 50% of women with endometrial carcinoma have a history of menorrhagia

4 Obesity, hypertension and diabetes – link is doubtful

5 Oestrogens – oestrogen replacement therapy given for the treatment of menopausal symptoms is associated with an increased risk of endometrial carcinoma

6 Endometrial polyps may increase risk of development of carcinoma

7 Commoner in higher socio-economic groups

Symptoms and signs:
1 Persistent watery discharge or a purulent discharge associated with a pyometra

2 Abnormal bleeding – usually postmenopausal. Thus endometrial carcinoma constitutes up to 40% of cases of postmenopausal bleeding in some series

3 In premenopausal women the bleeding is usually intermenstrual but may present as prolonged or heavy menstruation

4 Uterine colic may be present

5 Pain is a late symptom and evidence of extra-uterine metastases in lungs, spine or bones occurs late in the disease

Diagnosis:
1 Histological examination of uterine curettings

2 Cytological examination of smears from the posterior fornix is unreliable but endometrial aspiration (vabra suction aspiration) to obtain cells for cytological examination is more accurate but is associated with a 20% failure rate

Pathology: Endometrial cancer may arise from any region of the endometrium but begins most commonly in the fundus and the area of insertion of the Fallopian tubes. It is nearly always an adenocarcinoma of polypoid or nodular type. Early on these cancers grow towards the endometrial cavity and into the cervical canal. Most tumours are well differentiated and the more highly differentiated the tumour the better the prognosis. Adeno-acanthoma is a tumour with areas of squamous metaplasia. Adeno-squamous carcinomas contain about equal amounts of squamous and glandular elements. Both of these tumours are relatively rare.

Spread:
1 Local spread – tumour is in the endometrium itself and to a lesser extent into the myometrium. If the tumour invades the myometrium it may also involve the Fallopian tube, ovary and pelvic peritoneum

2 Lymphatic spread occurs to the para-aortic lymph nodes mainly via the ovarian vessels; inguinal and pelvic lymph nodes may also be involved

3 Once the tumour has spread to the cervix, lymphatic spread becomes identical to that of cervical carcinoma

4 Haematogenous spread is more frequent than in cervical carcinoma, because the endometrium is richly vascularized. Distant metastases occur to lungs, liver, bones and brain

Clinical staging: The different clinical stages of carcinoma of the corpus uteri (FIGO, 1979) are:

Stage 0:   Carcinoma *in-situ*. Histological findings suspicious of malignancy.

Stage I: Carcinoma confined to the corpus:

Stage 1a: The length of the uterine cavity is 8 cm or less

Stage 1b: The length of the uterine cavity is more than 8 cm. The Stage 1 cases should be sub-grouped with regard to the histological type of the adenocarcinoma as follows:

G1: highly differentiated adenomatous carcinoma

G2: differentiated adenomatous carcinoma with partly solid areas

G3: predominantly solid or entirely undifferentiated carcinoma

Stage II: Carcinoma has involved the corpus and the cervix

Stage III: Carcinoma has extended outside the uterus, but not outside the true pelvis

Stage IV: Carcinoma has extended outside the true pelvis or has obviously involved the mucosa of the bladder or rectum

Investigation:
1 Cytology is unreliable

2 Endometrial biopsy – using curette, brush or jet washings

3 Fractional curettage:
   a Cervical canal is scraped by a small curette before dilatation
   b After dilatation, the endometrium is then curetted
   The specimens are examined separately

Management:
1 Surgery – method of choice. Total hysterectomy with excision of a cuff of vagina and bilateral salpingo-oophorectomy produces about a 60% 5-year survival rate. Unfortunately in a few cases recurrences occur in the vault of the vagina

2 Surgery is usually performed in conjunction with preoperative intracavitary and vault radium applied usually by the

Stockholm or Manchester technique in some centres. Alternatively radiotherapy may be applied post-operatively – once the abdominal wound has healed and in the light of a full report by the pathologist. Postoperative radiation is also advised when histological examination of the excised uterus reveals that the cancer has penetrated deep into the myometrium and particularly if the serosal surface is involved

3 Radiotherapy alone – mainly used for inoperative cases

4 Hormone therapy: Hormone therapy provides a useful adjunct when other methods have failed. Endometrial cancer responds to high doses of progestagens in about one-third of cases. Most of these have well-differentiated carcinoma. Drugs commonly used include oral medroxyprogesterone acetate and intramuscular gestronol. Treatment is usually commenced at the time of initial curettage if malignant curettings are obtained

5 Cytotoxic therapy: Antimetabolites and alkylating agents have been used mainly for metastatic disease. Doxorubicin appears to be a promising drug especially in those tumours that show no response to progestagen therapy

Recurrence: These are found in about 20% of cases in the vagina, parametra, uterus and ovaries. Both recurrence and distant metastases have to be managed by hormone or cytotoxic therapy, and even these offer only temporary remission.

Prognosis: Five-year survival rate with treatment:
   Stage 0 – 100%
   Stage I – 90%
   Stage II – 75%
   Stage III – 15–40%
   Stage IV – 15–40%

SARCOMA OF THE UTERUS

Incidence: Around 2% of malignant uterine neoplasms are sarcomas

Types of uterine sarcoma:

Pathology:

1 Leiomyosarcoma:
   a Arising in leiomyofibroma
   b Arising in myometrium

2 Endometrial sarcoma:
   a Stromal sarcoma, to include some or all cases of stromal endometriosis
   b Mixed mesodermal-cell tumour, to include:
      i Sarcoma botyroides
      ii Osteosarcoma
      iii Chondrosarcoma
      iv Rhabdomyosarcoma
      v Fibromyosarcoma
      vi Carcinosarcoma

3 Vascular
   a Angiosarcoma
   b Haemangiopericytoma: some cases

4 Lymphosarcoma

Clinical features: Rapid growth may cause pain and a rapidly enlarging abdominal mass with uterine bleeding.

Management: Abdominal hysterectomy with bilateral salpingo-oophorectomy with possibly postoperative irradiation. Prognosis is poor.

MALIGNANT NEOPLASMS OF THE FALLOPIAN TUBES

Incidence: These are the least frequent genital carcinomas (0.1–0.8% cases). They usually occur beyond age 40.

Pathology: Adenocarcinomas, usually involving the isthmial portion of the tube. Usually well differentiated.

Spread:

1 Early direct spread, involving the peritoneum

2 Local and lymphatic spread to uterus, ovaries, bladder and rectum

3 Haematogenous spread is a late feature

Clinical features: Clinical symptoms are a late sign; diagnosis is in most instances accidentally at laparotomy.

Management: Identical as for ovarian tumours. Irradiation and chemotherapy may be tried. Prognosis is poor: 5-year survival is only 10%.

TUMOURS OF THE OVARY

Incidence: Ovarian cancer is now more common than invasive cervical cancer or endometrial cancer. Nearly 2% of all women may be expected to ultimately die of ovarian cancer. As much as 30% of ovarian tumours are, or eventually become, malignant.

Aetiology: Little is known:
1 More common in women in the fifth and sixth decades of life

2 More common in single than married women

3 Possibly more common amongst professional classes rather than lower socio-economic group

4 Reduced fertility and delayed childbearing may be predisposing factors

Symptoms and signs:
1 The topography of the ovaries virtually excludes early detection of ovarian tumours – symptoms occur late in the course of the disease

2 Pressure symptoms – due to primary growth and infiltration of surrounding tissues

3 Dissemination symptoms – from peritoneal implantations – causing abdominal distension due to ascites. Weight loss, anaemia and cachexia are late signs

4 Hormonal symptoms – postmenopausal vaginal bleeding, defeminization, masculinization, etc. These vary according to the type of tumour

5 Clinical signs of malignancy in an ovarian tumour are:
   a Fixity
   b Bilaterality – in about 25% of cases
   c Presence of ascites
   d Evidence of bowel involvement

6 Common symptoms related by sufferers from ovarian cancer are:
   a Abdominal pain and swelling
   b Abnormal vaginal bleeding

Acute presentation: This may result from the following complications of an ovarian tumour:

1 Torsion of the pedicle

2 Rupture

3 Incarceration

4 Infection

5 Haemorrhage

These may manifest as some of the features of an acute abdomen: abdominal pain and tenderness, nausea and vomiting, ileus and shock.

Pathology: The ovary comprises a wide variety of histological structures, which originate from different stem cells, which may proliferate and form a variety of tumours. These include:

1 The mesenchymal stroma around the cellular part of the follicle including granulosa and thecal cells

2 The germinal epithelium on the surface of the ovary

3 Androgen-producing hilar cells

4 Pluripotential remnants of the embryonic gonadal anlage, e.g. mesenchyme and coelomic epithelium

A generally accepted classification is not available, and distinction between benign and malignant tumours is unclear, as many benign tumours are potentially malignant.

Histological classification of the common
primary tumours of the ovary (FIGO, 1979):

1 Serous cystomas:
   a Serous benign cystadenomas
   b Serous cystadenomas with proliferating activity of the
     epithelial cells and nuclear abnormalities but with no
     infiltrative destructive growth (low potential malig-
     nancy)
   c Serous cystadenocarcinomas

2 Mucinous cystomas:
   a Mucinous benign cystadenomas
   b Mucinous cystadenomas with proliferating activity
     of the epithelial cells and nuclear abnormalities but
     with no infiltrative destructive growth (low potential
     malignancy)
   c Mucinous cystadenocarcinomas

3 Endometroid tumours (similar to adenocarcinomas in the
  endometrium):
   a Endometroid benign cysts
   b Endometroid tumours with proliferating activity of
     the epithelial cells and nuclear abnormalities but
     with no infiltrative destructive growth (low potential
     malignancy)
   c Endometroid adenocarcinomas

4 Mesonephric tumours:
   a Benign mesonephric tumours
   b Mesonephric tumours with proliferating activity of the
     epithelial cells and nuclear abnormalities but with no
     infiltrative destructive growth (low potential malig-
     nancy)
   c Mesonephric cystadenocarcinomas

5 Concomitant carcinoma, unclassified carcinoma (tumours
  which cannot be allotted to one of the groups 1, 2, 3 or 4).

Spread: Malignant ovarian tumours spread to para-aortic,
mediastinal and supraclavicular lymph nodes. Metastases occur
late to distant organs, chiefly the lungs and liver.

Clinical staging:

The FIGO (1979) stage grouping classification is based on findings at clinical examination and surgical exploration:

Stage 1: Growth limited to the ovaries

    Stage 1a: Growth limited to one ovary; no ascites

        (i)   Capsule not ruptured

        (ii) Capsule ruptured

    Stage 1b: Growth limited to both ovaries; no ascites

        (i)   Capsule not ruptured

        (ii) Capsule ruptured

    Stage 1c: Growth limited to one or both ovaries; ascites present with malignant cells in the fluid

        (i)   Capsule not ruptured

        (ii) Capsule ruptured

Stage II: Growth involving one or both ovaries with pelvic extension

    Stage IIa: Extension and/or metastases to the uterus and/or tubes and/or other ovary

    Stage IIb: Extension to other pelvic tissues

Stage III: Growth involving one or both ovaries with widespread intraperitoneal metastases

Stage IV: Growth involving one or both ovaries with distant metastases

Special category: Unexplored cases which are thought to be ovarian carcinoma

In 1968, the classification based on the TNM system was suggested. It has the advantage that the extent of the tumour is stated more precisely:

    TP: extension of the primary tumour

    TPI: tumour mobile and limited to one ovary

    TPII: tumour mobile but involving both ovaries

    TPIII: uterus and/or Fallopian tubes involved

    TPIV: other neighbouring anatomical structures involved

    N: regional lymph nodes

    NX: in ovarian carcinoma the pelvic nodes cannot be evaluated

NX+ and NX−:   denote positive or negative lymph nodes at laparotomy
M: distant metastases
MO: no distant metastases
MI: metastases present:
    a. in the pelvis
    b. in the peritoneal cavity
    c. outside the peritoneal cavity

Management

1 General: All ovarian tumours should be removed surgically; this is also effective prophylaxis in view of the malignant potential of many 'benign' lesions of the ovaries.

2 Benign tumours: Unilateral oophorectomy will suffice for women in the reproductive age; small areas of normal functioning ovary may be preserved after removing cysts. However, inspecting and obtaining a biopsy of the contralateral ovary is preferable for most tumours for histological examination

3 Bilateral tumours, or postmenopausal women: Bilateral salpingo-oophorectomy and hysterectomy

4 Malignant tumours of the ovary: Treatment is primarily surgical − total abdominal hysterectomy and bilateral salpingo-oophorectomy or bilateral salpingo-oophorectomy alone. The aim of surgery is to remove the maximum amount of tumour at the initial surgical procedure. Subsequent therapy may involve:
    a Pelvic irradiation
    b Pelvic and abdominal irradiation
    c Chemotherapy − single agent or combined agents − chlorambucil, cyclophosphamide, melphalan and triethylene thiophosphoramide are the most commonly used alkylating agents. Adriamycin, 5-fluorouracil, methotrexate, vinblastine and vincristine are also used
    d Combined irradiation and chemotherapy, synchronous or sequentially

Prognosis: The overall 5-year survival rate from cancer of the ovary in treated cases is only about 10% depending on the stage and histology of the tumour. Over 50% of patients have an advanced stage of growth at time of presentation.

6

Urological Problems

Up to 20% of gynaecological patients may complain of associated urinary symptoms.

KIDNEY AND URETER

1   Renal aplasia:
    a   Clinical features: May be asymptomatic. Cystoscopy reveals only one ureteric orifice. Intravenous pyelography reveals the absence of one kidney and ureter
    b   Management: Renal function is normal. Vigorous treatment of infections of urinary tract, especially during pregnancy

2   Pelvic kidney:
    a   Clinical features: May be mistaken for an ovarian tumour on palpation
    b   Management: Treatment unnecessary

3   Ureteric variations:
    a   Examples include complete or partial duplication of the ureter on one or both sides
    b   Generally symptomless, but may be accidentally damaged during surgery

LOWER URINARY TRACT

Asymptomatic bacteriuria

Predisposing factors:
1 Pregnancy
2 Puerperium
3 Prolapse
4 Postoperatively, especially if catheterization was used
5 After radiotherapy
5 Genital carcinoma

Symptoms: Asymptomatic
Diagnosis: Bacteriological examination of midstream specimen of urine
Management: Antibiotics, e.g. ampicillin 1 g t.d.s. Choice of antibiotics may be modified with knowledge of bacteriological results.

Cystitis

Aetiology:
1 Ascending infections
2 Usually *Escherichia coli*, *Proteus* or *Neisseria gonorrhoeae* involved

Predisposing factors:
1 Postmenopausal
2 Catheterization
3 Previous urinary tract surgery
4 Traumatic delivery of infant

Clinical features:
1 Intermittent pain over the pubic area
2 Frequency and dysuria
3 Cloudy urine

Investigation:
1 Microscopy and culture of suprapubically obtained urine specimen

2 If recurrent or persistent infection, pelvic examination to exclude extravesical cancer and cystoscopy may be needed

Management:
1 Bed rest and local heat

2 Adequate fluid intake

3 Antibiotic treatment, e.g. ampicillin 1 g t.d.s. (may be modified in the light of bacteriological results)

Neoplasms of the bladder

1 Benign: These include papillomas, myomas, fibromas or dermoid cysts. They are nevertheless considered as premalignant and require treatment

2 Malignant:
  a Primary tumours:
    i Often originate from a papilloma
    ii Causes haematuria and tenesmus
    iii May cause urinary obstruction
  b Secondary tumours
    i May originate from primaries in cervix, endometrium or vagina
    ii May result in bladder perforation or formation of fistulae

3 Endometriosis: May cause haematuria during menstruation

Urethritis

Aetiology: Usually gonococcal

Clinical features:
1 Severe dysuria

2 Reddened urethral orifice with pus extruded

Management: Antibiotic management

Urethral caruncle

A small, reddened, sensitive fleshy excrescence at the urethral meatus. Commonest beyond the menopause.

Pathophysiology:
1 Most caruncles are due to an eversion (ectropion) of the urethra or infection of the urinary meatus

2 A small percentage of caruncles may arise from vascular abnormalities or malignant or benign neoplasms

Clinical features:
1 History:
   a Dysuria, frequency and urgency
   b Local discomfort, dyspareunia
   c Bleeding and leukorrhoea

2 Examination:
   a Small, red mass protruding from the urethral meatus
   b May bleed or be tender

Investigation:
1 Bacteriological:
   a Urinary microscopy and culture
   b Microscopy and culture of smear
   c Dark-field examination if syphilis suspected

2 Smears for cytology and biopsy, to exclude malignant change

Management:
1 General:
   a Sedation and analgesia to relieve distress
   b Manage any associated infection
   c Consider diethylstilboestrol vaginal suppositories if postmenopausal patient

2 Surgical:
   a If tests imply benign lesions: diathermy or cryosurgery under anaesthesia
   b If malignant: radical resection or radical excision

Urethral diverticula

Pathophysiology:
1 These are sacculations in the urethral wall that may result from:
   a Congenital dilation of Wolffian duct remnants
   b Surgical injury
   c Infection of the paraurethral glands resulting in rupture into the urethra

2 Symptoms result from either the sacculation itself, or intercurrent infection, or calculus formation in the diverticulum

Predisposing factors:
1 40–50 years old

2 Multiparous patients

Clinical features:
1 History:
   a Onset may be acute or chronic – most commonly chronic infections with periodic acute exacerbations
   b Features of intermittent infection – malaise, chills, pyrexia
   c Urinary symptoms:
      i Dysuria, urgency, frequency, nocturia
      ii Post-voiding dribbling of urine
   d Local infection:
      i Discharge at meatus may be purulent or bloody
      ii Vaginal pain

2 Examination: Tender anterior vaginal fullness or soft mass on digital examination

Investigation:
1 Radio-opaque contrast studies of the urethra

2 Urethral sound may demonstrate the diverticulum

Differential diagnosis:
1 Urethritis

2 Urethrocoele

3   Gartner's duct cysts

4   Tumours on or near the urethra

Management:
1   Management of an acute inflammatory episode:
   a   Broad-spectrum antibiotics
   b   Adequate analgesia

2   Definitive treatment: Transvaginal diverticulectomy

INCONTINENCE OF URINE

Physiology:
1   Urethra: contains several components that contribute to urinary continence
   a   Smooth muscle in mucosa and striated bulbocavernosus and ischiocavernosus muscles in anterior third
   b   Muscle layer underlying corrugated mucosa in middle third
   c   Urethral wall similar to bladder in upper third

2   Pubo-urethral 'sling':
   a   Sling-shaped bundles of connective fascia and smooth muscle pull the urethra at its posterior third and preserve a vesico-urethral angle of 100° at rest
   b   On micturition, the pelvic diaphragm relaxes and the urethra descends. The vesico-urethral angle becomes less acute. The trigonal area of the bladder fills with urine. This contracts and increases the intravesical pressure, and urine is released
   c   At the end of micturition, the pelvic diaphragm contracts. This pulls the urethra forwards, and restores the vesico-urethral angle

Stress incontinence

Definition: this is the involuntary escape of a small amount of urine when the patient strains, coughs or laughs.

Clinical features:
1   History: stress incontinence may be graded as follows:

   a Slight loss of urine during coughing, laughing or
severe effort

   b Spontaneous micturition during running, carrying,
climbing stairs or light physical work

   c Incontinence when standing

2 Examination:

In most cases there is:

   a Moderate to severe cystocoele

   b Blunt urethrovesical angle

   c On palpating the urethra, the urethra is easily displaced and has lost its firm outlines

Other rarer causes of stress incontinence must be excluded.

Less common causes of stress incontinence:

1 Detrusor instability:

   a Local bladder pathology:
- **i** Infection
- **ii** Neoplasm
- **iii** Diverticulum
- **iv** Trigonitis

   b Lower motor neurone lesions:
- **i** Spina bifida occulta
- **ii** Neoplasms
- **iii** Prolapsed disc
- **iv** Pelvic surgery

   c Upper motor neurone lesions:
- **i** Cerebrovascular accident
- **ii** Multiple sclerosis
- **iii** Spinal cord transection

   d Large bowel pathology
- **i** Diverticular disease
- **ii** Tumours

2 Sphincter pathology:

   a Obstetric trauma
- **i** Prolonged or precipitate labour
- **ii** Large baby
- **iii** Forceps delivery before full cervical dilatation

   b Postoperative

   c Congenital

3 Urethral narrowing:
   a Previous gynaecological surgery
   b Recurrent infection
   c Obstetric trauma
   d Senile vulvitis

Urge incontinence

Definitions: intense desire to micturate even when the bladder hardly is filled with urine

Causes:
1 Psychogenic – 80% of cases

2 Cystitis

3 Urinary tract calculi

4 Tumours

5 Tuberculosis

6 Prolapse, especially cystocoele

Overflow incontinence

Overflow incontinence may reflect the presence of urinary tract outflow obstruction. It may lead to acute retention.

Causes:
1 Factors outside the urethral wall:
   a Pelvic masses
   b Fibroids
   c Kinking of the urethra in prolapse

2 Factors in the urethral wall:
   a Stricture
   b Tumour

3 Factors in the urethral lumen: calculi

4 Neurological causes

Management of stress incontinence

1 Prevention during and after pregnancy:
  a Antepartum and postpartum exercises to strengthen pelvic floor structures
  b Avoidance of obstetric trauma in labour

2 Medical management
  a Indications:
    i Small degree of incontinence
    ii Co-operative patient
  b General measures:
    i Manage obesity, cough and associated medical disorders
    ii Physiotherapy – faradism to pelvic muscles, pelvic exercises

3 Surgical management:
  a Indications:
    i Presence of cystocoele or utero-vaginal prolapse
    ii Failure of medical treatment
  b Operations are directed at:
    i Restoring the urethrovesical angle
    ii Tightening the musculofascial structures around the urethrovesical junction
    iii Correcting any cystocoele or prolapse present
  c Operations include:
    i Kelly operation: Urethroplasty and anterior colporrhaphy. Plication of the pubovesical fascia and of other pelvic floor supports
    ii Marshall–Marchetti operation: Urethrocystopexy drawing the bladder and urethra upward and forward
    iii Millin operation: Sling operation carrying a band of fascia or nylon beneath the posterior urethra for support of the urethrovesical angle

4 Postoperative care:
  a Indwelling catheter for 4–5 days to deal with postoperative retention

b Vigorous treatment of any urinary tract infections

Prognosis: Overall success rate with first vaginal operation is 50% provided the initial abnormality was not large.

Continuous incontinence

This may be caused by:

1 Congenital urinary tract abnormality

2 Fistula formation

Causes of urinary fistulas:

1 Malignant tumours

2 Postoperative

3 Difficult or traumatic labour

4 Crohn's disease

5 Ulcerative colitis

Investigation of incontinence

1 Urine microscopy and culture to examine for infection

2 Methylene blue test to assess for degree of cystocoele retention of urine:
   a Bladder filled with a diluted methylene blue
   b Patient then micturates until bladder feels empty
   c Patient then walks about and bears down and the resulting micturition effect estimates the degree of cystocoele retention

3 Radiological investigations:
   a Lateral resting and voiding cysto-urethrograms to examine the urethrovesical angle
   b Antero-posterior cysto-urethrograms – confirmatory evidence

4 Special tests:
   a Urethroscopy or cystoscopy to exclude developmental or traumatic abnormalities
   b Cystometry to investigate bladder pressure – volume relationships

7

Infertility

Definitions

1 Infertility: The absolute absence of the ability to conceive

2 Subfertility: The failure to conceive after a year of normal coitus

3 Primary infertility: No previous pregnancy in history

4 Secondary infertility: Previous pregnancy did occur

Prevalence

About 8% of marriages are childless. A further 10–12% of married couples have only 1–2 children but desired more.

Aetiology

1 Sole cause in the female	30%
2 Sole cause in the male	30%
3 Combined male and female causes	30%
4 No recognizable cause	10%

Sole cause in the female:
1 Defective ovulation – 20%:
    a Permanent failure to produce ova – associated with amenorrhoea, e.g. intersex states
    b Temporary failure to produce ova – associated with oligomenorrhoea or secondary amenorrhoea

141

2 Mechanical blockage of oviducts:
    a Adhesions – post-appendicitis and peritonitis
    b Previous salpingitis – post-abortal, puerperal, tuberculosis, gonococcal
    c Congenital

3 Uterine factors:
    a Congenital abnormalities, e.g. bicornuate uterus
    b Tumours – fibroids
    c Endometriosis
    d Endometritis
    e Retroverted uterus

4 Cervical factors:
    a Hostile cervical mucus
    b Cervical disease
    c History of cervical amputation

5 Pregnancy wastage – spontaneous abortions

6 Coital errors:
    a Infrequent or no intercourse
    b Use of lubricants
    c Urethral coitus

Sole cause in the male:
1 Failure of adequate spermatogenesis:
    a Chromosomal abnormalities such as Klinefelter syndrome (**XXY**)
    b Maldescended or undescended testes
    c Damage to the testes – injury, orchitis (especially mumps, but also caused by tuberculosis or venereal disease) irradiation, previous surgery and tumours
    d Intercurrent diseases such as diabetes, thyroid or adrenal dysfunction
    e Excessive heat applied to the scrotum, for example tight clothing and varicocoeles

2 Obstructed seminal ducts
    a Congenital absence of ducts
    b Previous epididymitis (tuberculous or gonococcal)

    c Trauma due to accident or injury to the vas deferens during hernia repair

3 Failure to deposit sperm in vagina
    a Impotence
    b Premature ejaculation
    c Hypospadias or phimosis
    d Retrograde ejaculation into the bladder especially after prostatectomy

INVESTIGATION OF THE INFERTILE COUPLE

From the first visit the couple must be seen together. A full history must be taken from both partners.

History

1 Age, occupations and nationality, past medical, surgical and family histories

2 Marital history:
    a Length of marriage
    b Length of time trying to become pregnant
    c Frequency and satisfaction of coitus

3 Menstrual history:
    a Age of menarche
    b Nature and frequency of menstruation
    c Date of last menstrual period
    d Use of contraceptives; methods and length of use
    e Previous fertility investigations

4 Obstetric history:
    a Full details of previous pregnancies and deliveries
    b Previous abortions

5 History of previous sexually transmitted diseases

General physical examination

Male:
1 General physique and secondary sexual characteristics

2 Examine penis and scrotum for congenital abnormalities

3 Note size, consistency and position of testes

4 Palpate for presence of vas deferens bilaterally

5 Exclude varicocoele, hydrocoele and herniae

6 Assess prostate and seminal vesicles by rectal examination

7 Examine heart, lungs and abdomen

Female:
1 General physique and secondary sexual characteristics – note presence of hirsutism, exophthalmos, thyroid enlargement, pigmentation, striae

2 General examination of the heart, lungs, blood pressure, breasts (look for lumps and galactorrhoea) and abdomen

3 Pelvic examination:
  a Is vulva normally developed or immature and are the hymen and vagina properly stretched?
  b Presence of vaginismus or congenital abnormality
  c Vaginal and pelvic infection
  d State of the cervix
  e Position, size and mobility of the uterus
  f Presence of ovarian enlargement or any adnexal masses
  g Evidence of endometriosis

First visit

If possible the patient's first appointment should be made at about the time of ovulation. If possible she should be told to have intercourse the night before her appointment so that a postcoital test (see below) may be done
  1 General laboratory investigations:
    a Both partners:
      i Full blood count and ESR
      ii Blood group
      iii Serological tests for syphilis
      iv Chest X-ray if indicated (lung disease or history of tuberculosis)

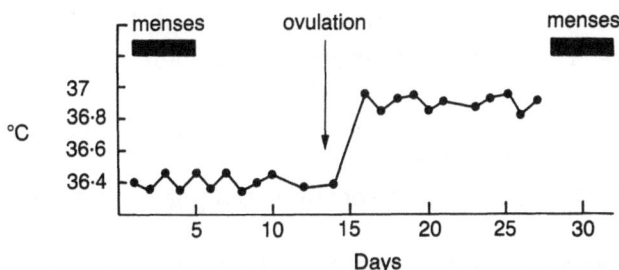

Figure 7.1 *Body temperature chart showing that ovulation was presumed to have occurred on day 14 of the cycle*

b Female:
  i Rubella antibodies (immunize if negative)
  ii Urinalysis and urine microscopy and bacteriology

2 Cervical cytology

3 Confirmation of ovulation:
  a The patient is instructed to keep a basal body temperature record, measuring her body temperature upon waking each morning. Following ovulation, there is a rise of temperature of 0.2–0.6°C for 14 days (see Figure 7.1)
  b Ovulation has also probably occurred if a plasma progesterone measurement 7 days after the temperature change is over 10 ng/ml

4 Postcoital (Sims–Huhner) test is best performed within 12 hours of intercourse after abstinence of 3–5 days, preferably at mid-cycle. A drop of cervical mucus is placed on a slide and examined under the microscope. Ovulatory mucus should be clear and thin and contain few cells and many spermatozoa. It is necessary to see a few sperm (5 or more) in progressive motion per high power field over at least twelve such fields for the test to be positive. Results may be:
  a Normal or positive
  b Reduced sperm count – 1–5 sperm with good motility
  c No sperms
  d Dead sperms
One semi-quantitative scoring system for assessing the postcoital test is the Swyer score:

Score 0: No spermatozoa seen in high power field

Score 1: Around 5 spermatozoa per high power field; around 10–20% motile

Score 2: 10–30 spermatozoa per high power field; up to 50% motile

Score 3: 40 or more spermatozoa per high power field. All motile

If the postcoital test is negative it should be repeated several times

5 Arrange seminal analysis

6 If history suggestive of tubal damage or endometriosis discuss investigation of tubal patency

Second visit

1 Review basal body temperature chart – at least 30% will be abnormal
  a Monophasic chart – anovulation (check that patient is recording temperature properly)
  b Slow rise in temperature – defective ovulation
  c Short elevation of temperature – failure of corpus luteum

2 Review seminal analysis: The specimen should be produced by masturbation after 5 days abstinence from intercourse, and should be examined within 4 hours of collection:
  a Volume – should be between 2.0 and 6.0 ml
  b Viscosity – liquefaction should be complete within 30 minutes
  c pH – should be between 7.2 and 7.7
  d Sperm count – the density should be between 20 and 250 million/ml. About 50% of men have sperm counts between 40 and 60 million/ml
  e Form – no more than 30% of sperm should be abnormal
  f Motility – greater than 60% of sperm if examined within 4 hours
  g Viability – no more than 50% of sperm should be dead
  h Leukocytes – ? prostatitis
  i Agglutination – caused by autoimmune antibodies

3   The Kürzrok–Miller test is indicated in the event of re-
    peated negative postcoital tests. Invasion of ovulatory cer-
    vical mucus by part of the sperm sample is observed under a
    high-power microscope for 15 minutes. The possible out-
    comes are:

    a   A normal result occurs when normal spermatozoa im-
        mediately start to penetrate cervical mucus, and con-
        tinue to remain active within it
    b   Azoospermia, oligospermia or low motility will have
        previously been detected on seminal fluid analysis. No
        sperm, dead sperms or low numbers of sperm will be
        seen on the Kürzrok–Miller test
    c   Sperm autoantibodies will also have been detected on
        seminal fluid analysis. Again no sperm invasion will
        be detected
    d   Hostile mucus. Spermatozoa only make a limited pen-
        etration into the mucus and then become immobilized
        and die; the same sperm should show normal pen-
        etration through cervical mucus of a healthy 'control'
    e   Sperm antibodies in the cervical mucus. Spermatozoa
        penetrate the mucus but then become agglutinated
        and immobilized

4   Repeat postcoital test

5   If necessary arrange for further tests of ovulation (see
    below)

Test to determine ovulation

1   Basal body temperature chart

2   Cervical mucus ferning – at ovulation stretching the thin
    cervical mucus produces long threads (Spinnbarkeit). If this
    mucus is allowed to dry on a slide sodium chloride crystals
    are precipitated and produce a fern-leaf pattern. After ovu-
    lation cervical mucus is thicker and the ferning pattern is
    not easily seen

3   Endometrial biopsy – a biopsy sample is taken in the week
    preceding menstruation. An atrophic endometrium indi-

cates an inadequate secretion of oestrogen and or progesterone. The test is not in common use

4 Hormone assays:
   a Plasma progesterone – easy test in common use. Two samples are needed – one early (about day 18) and the other later in the luteal phase (about day 24) in a 28 day cycle.
      i Levels about 16 nmol/1 in the luteal phase suggest normal ovulation
      ii A slow rise in progesterone levels may indicate defective ovulation
      iii A rapid fall in progesterone levels may indicate a defective corpus luteum
   b Prolactin – upper limits vary from laboratory to laboratory but levels over 500 mU/litre are regarded as abnormal
   c 'Dynamic' endocrine tests:
      i LHRH test – I.V. administration of 25–100 µg of gonadotrophin-releasing hormone GnRH (LH-RH) should cause a rapid increase in serum LH and a small rise in FSH
      ii Clomiphene stimulation test: Clomiphene is given orally for 7 days starting on day 3 of the menstrual cycle if menstruating. LH, oestrogen and progesterone levels are examined over the next two to three weeks. A normal response is indicated by a rise in plasma LH during the test period and a peak in plasma progesterone post-ovulatory
      iii Human menopausal gonadotrophin (HMG) stimulation: HMG contains equal amounts of FSH and LH and is given on three alternate days (1, 3, and 5) followed by a single injection of HCG on day 8. The response is measured by estimating the increase in total urinary oestrogen in a 24 hour specimen.
These tests provide a measure of the functional integrity of the hypothalamic-pituitary-ovarian axis

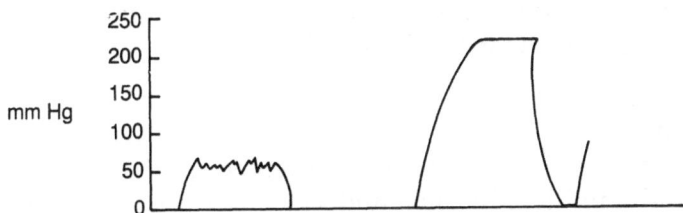

Figure 7.2 *Tracings of insufflation of the Fallopian tubes*

Tests of tubal patency

1 Tubal insufflation (Rubin's test): An easy test best performed on a conscious woman about mid-cycle. Carbon dioxide is introduced into the cervical canal and the pressure
   tracing is recorded on a kymograph. The pressure should
   not be allowed to rise above 200 mmHg (Figure 7.2)
   a Positive result
      i Low pressure reading
      ii Sound of gas passing through tubes on ausculta-
         tion
      iii Shoulder pain caused by subdiaphragmatic irri-
          tation
   b Negative result
      i Tubal spasm – give an antispasmodic agent (amyl
        nitrite)
      ii No gas heard and gas pressure rises
Contraindications:
   a Uterine haemorrhage
   b Pelvic infection
   c Purulent vaginal discharge
The technique is rarely used since it provides no indication as to
whether only one tube is patent or the extent of tubal damage.

2 Hysterosalpingography: A radiological examination after
   intrauterine injection of radio-opaque medium. This shows
   the size and shape of the uterine cavity, the condition of the
   tubes and whether there is overflow through the ends of the
   tubes. The test is best performed in the first half of the cycle

so as not to interfere with a possible pregnancy. Contraindications are as for tubal insufflation

3 Laparoscopy: This technique usually requires general anaesthesia and involves introducing carbon dioxide into the abdominal cavity. A laparoscope is introduced through a subumbilical incision and the pelvic organs can be visualized. A corpus luteum in an ovary is positive evidence of ovulation. If necessary an ovarian biopsy can be taken and small peritoneal adhesions can be divided. Simultaneous injection of a dye into the cervical canal demonstrates the tubes. Laparoscopy is a common procedure since in one simple operation ovulation can be confirmed, pelvic pathology can be excluded and tubal patency can be shown
Note: Laparoscopy is contraindicated with a history of previous abdominal operations or peritonitis owing to the danger of adhesions attached to the surgical scar. It is contraindicated in gross obesity or severe cardiac or pulmonary disease (see Chapter 1)

MANAGEMENT OF INFERTILITY IN THE FEMALE

Treatment of ovulatory failure

Classification:
Group A   Hypothalamo-pituitary failure:
   Amenorrhoeic
   No endogenous oestrogen
   Normal or low FSH levels
   Normal prolactin levels
Group B   Hypothalamo-pituitary dysfunction:
   Menstrual disturbances including luteal phase deficiency, anovulation or amenorrhoea
   Normal oestrogen, prolactin and FSH levels
Group C   Ovarian failure:
   Amenorrhoeic
   No endogenous oestrogen
   High FSH levels
   Normal prolactin levels
Group D   Genital tract disorder:
   Amenorrhoeic

Do not respond by withdrawal bleeding to exogenous oestrogen

Group E Hyperprolactinaemia. This will result in: Menstrual disturbances

Elevated prolactin levels

Where due to a prolactin-secreting tumour, the pituitary may be enlarged

Amenorrhoea is discussed further in Chapter 3

Hyperprolactinaemia is associated with amenorrhoea with or without galactorrhoea (abnormal milk secretion). In both man and woman hyperprolactinaemia also causes a loss of libido. High prolactin levels inhibit the preovulatory LH surge and contrary to earlier opinion has little effect on ovarian responsiveness to FSH and LH. The most sinister cause of hyperprolactinaemia is a prolactin-secreting pituitary adenoma and investigations are needed to eliminate this diagnosis.

Causes of hyperprolactinaemia:

1 Stress

2 Pregnancy and during the first few months of breast feeding

3 Oral contraceptives – oestrogens

4 Psychoactive drugs e.g. chlorpromazine and haloperidol (works by blocking dopamine receptors in the anterior pituitary

5 Metoclopramide treatment

6 Polycystic ovary syndrome

7 Pituitary tumours – adenomas or microadenomas

Prolactin secretion from the lactotroph cells in the anterior pituitary is under the control of a release-inhibiting factor called prolactin-inhibiting factor (PIF) which has been identified as dopamine.

Bromocriptine therapy

The ergot alkaloid bromocriptine, a dopamine agonist, is the drug of choice for suppressing excessive prolactin secretion. Inappropriate hyperprolactinaemia is thought to account for about 20% of patients with amenorrhoea. It is given at a dose of 2.5 mg twice a day (with food) for 14 days. Because of side effects bromocriptine is best given gradually or stepwise starting with 1.25 mg daily and building up to 2.5 mg twice daily over 12 days. Side effects include:

1 Nausea and vomiting

2 Postural hypotension

3 Dizziness and headaches

The response is assessed by monitoring prolactin and progesterone concentrations in the third week of the ensuing menstrual cycles. A prolactin-secreting pituitary tumour must be excluded in all patients with hyperprolactinaemia. This is usually done by an X-ray of the pituitary fossa and if abnormal followed up by air encephalography or computerized tomography. A danger therefore of bromocriptine therapy is that it may be given to patients with unrecognized pituitary tumours though some recent evidence has shown that bromocriptine may cause prolactinomas to shrink

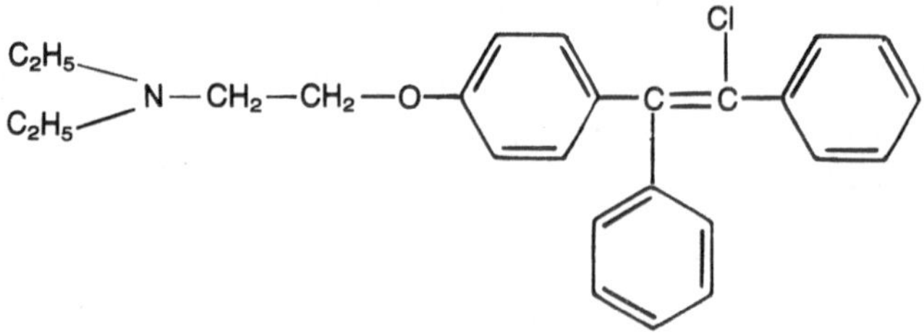

Figure 7.3 *Structural formula of clomiphene citrate*

Clomiphene therapy

Action of clomiphene (Clomid) (Figure 7.3): Clomiphene is a non-steroidal anti-oestrogenic compound which acts by blocking the long loop inhibitory feedback effect of oestrogen on the hypothalamus. The hypothalamus is then unable to detect the actual level of circulating oestrogen and it responds by releasing luteinizing hormone releasing hormone with a subsequent increase in follicular stimulating hormone (FSH) secretion. FSH stimulates follicular development and ovulation occurs in about 70% of cases.

Indications for clomiphene therapy: Clomiphene is used in Groups A and B of the above classification, that is in cases of hypothalamo-pituitary failure and dysfunction.

Regime: Clomiphene is given by mouth, the starting dose is 50 mg daily for five days from days 2–6 or from days 5–9 of the cycle. This may be repeated for two cycles while the basal body temperature is recorded. If ovulation has not occurred the course can be repeated with 100 mg or 150 mg daily for five days but the maximum dose is seldom used.

Side effects of clomiphene:

1 Anti-oestrogenic effect may make cervix hostile and may prevent pregnancy despite ovulation

2 Ovarian hyperstimulation syndrome – cystic enlarged ovaries, ascites and occasionally intra-abdominal bleeding. This is rare

3 Multiple pregnancies – incidence increased by eightfold (90% twins, 10% triplets etc.)

4 Hot flushes

5 Blurring of vision

6 Nausea

7 Breast discomfort

Results: 70% of anovular women ovulate with clomiphene therapy. 50% of these become pregnant but about 20–25% of all pregnancies end in abortion. If anovulation does not respond to clomiphene alone then gonadotrophins may be used

Gonadotrophin therapy

FSH is given first to produce follicles in the ovary and then LH is given to induce ovulation and the formation of the corpus luteum. Human menopausal gonadotrophin (HMG) containing both FSH and LH is used to develop the follicle and human chorionic gonadotrophin (HCG) containing LH is used to induce ovulation. Monitoring of urinary oestrogen excretion and plasma oestradiol levels during treatment is obligatory. Treatment is expensive and can only be carried out in well-equipped centres. Indications:

1 Hypothalamo-pituitary failure and dysfunction (Groups A and B)

2 Failure to respond to clomiphene

Side effects:

1 Hyperstimulation syndrome

2 Multiple pregnancies – varies between 12–45% of all pregnancies. When control is good 75% of multiple pregnancies are twins

Treatment of tubal occlusion

Tubal pathology is responsible for about 30% of female infertility. The current epidemic of venereal disease suggests that this figure may be even higher in the future. Successful tubal surgery depends on where the tube is occluded and the degree of damage to the ciliated lining of the tube. The types of operation that are available include:

1 Lysis of adhesions – salpingolysis – restores hormonal motility to the tubes

2 Salpingostomy or fimbrioplasty – to relieve fimbrial blockage

3 Tubal implantation – performed for blockage of the inner portion of the tube

4 Tubal reanastomosis – used to reverse sterilization operations

Further details of these operations can be found in Chapter 9.

The overall pregnancy rates following tubal surgery are less than 20% and the couple should be warned of this gloomy prognosis. Where the tubes are absent or irremediably blocked husband's sperm can be added in the laboratory to ova removed by laparoscopy. *In vitro* fertilization occurs and the developing blastocyst is implanted in the uterus (test-tube baby). At the present time a few normal births have resulted from this technique.

MANAGEMENT OF INFERTILITY IN THE MALE

Azoospermia

Lack of spermatozoa in the semen or deficiency in their functional activity.

Treatment of azoospermia depends on the underying cause. A blocked vas may be treated by reconstructive surgery but subsequent pregnancy is very low. Hypogonadotrophic hypogonadism is the only treatable cause of azoospermia. HMG is given daily for 110 days and HCG is given every week.

Oligospermia

The treatment of oligospermia is disappointing. Clomiphene 50 mg daily for 10 weeks or mesterolone 25 mg t.d.s. may produce an increase in sperm count but rarely enough.

Wearing loose underpants and avoiding hot baths may also raise the sperm count. Varicocoeles occur in about 25% of infertile males. Surgical treatment is often very successful.

Further tests:
1 Blood tests:
   a Full blood count and ESR
   b Thyroid profile – T4, T3 and TSH
   c FSH and LH levels
   d Testosterone level

2 Testicular biopsy: Not a commonly used test but some urologists combine this procedure with an exploration of the scrotum in cases where azoospermia is thought to be due to an

obstruction in the vas. In cases of severe oligospermia the biopsy may show generalized dysplasia of spermatogenesis

3 Vasogram: Demonstrates radiologically the level of obstruction. Not in common use

Ejaculatory failure

1 Exclude neurological causes for e.g. diabetes and multiple sclerosis

2 Exclude retrograde ejaculation – examine urine postcoitally; psychotherapy for psychogenic causes

Artificial insemination

AIH means the use of husband's semen and AID means the use of semen from a donor

1 AIH: Indications:

    a In cases of male impotence that are resistant to psychotherapy

    b In cases of premature or retrograde ejaculation

    c Moderate oligospermia

A fresh specimen of split ejaculate (sperm 'split' into two portions – the first portion is sperm rich and the second portion originates from the seminal vesicles) is usually used. The success rate for thawed frozen concentrated specimens from low density semen is low

2 AID: The combined problems of male infertility and difficulties in adoption have increased the demand for AID. Semen is selected from anonymous healthy donors and intracervical inseminations are carried out at least twice in the peri-ovular period. Semen may be frozen for long periods in 'sperm-banks'

    a Great care must be taken to protect the identity of the donor. Where possible the physical characteristics of the donor should resemble the husband

    b Donor inseminations do not guarantee pregnancy – the success rate is about 70%

    c The couple must be prepared by long discussions so

that they understand the full implications of the technique

d   Prior investigation of female infertility is essential

e   Single women and lesbian couples have requested AID

8

Contraception and Sterilization

Contraception is the prevention of conception in the uterus. A wide variety of means of contraception are now available. Each method can conveniently be considered under the following headings:

1 Effectiveness

2 Cost

3 Acceptability

4 Availability

5 Contraindications

6 Side effects

Effectiveness is usually measured by the Pearl Index which expresses a failure rate per 100 woman years (100 wy):

$$\frac{\text{Number of unintended pregnancies in population}}{\text{Total months of exposure to pregnancy}} \times 100$$

An 'ideal' contraceptive should in addition be safe, and not interfere with the enjoyment of intercourse by either party.

A woman seeking contraceptive advice should be asked for a detailed history. On physical examination, reproductive tract abnormalities and certain medical conditions should be excluded. The patient should be carefully instructed about the advantages, disadvantages and use of appropriate methods of contraception. It may be necessary to give her simple written or printed instructions to follow.

Use of the different available methods of contraception
(Family Planning Association estimate in 1979)

Pill	28%
Sheath	25%
No method or withdrawal	17%
Sterilized	5%
Intrauterine device	5%
Cap	3%
Rhythm method	2%
(Infertile	10%)
(Pregnant or trying for a baby)	5%)

CONTRACEPTIVE METHODS USED BY BOTH PARTNERS

Rhythm method ('safe period')

This is the avoidance of coitus around the time of ovulation. For example, in a 28 day cycle, ovulation takes place on about day 14. Allowing for the survival time for both the ovum and sperm – intercourse should not take place from about day 9 to about day 19 of the cycle.

Calculation of the safe period:

1 Method (1): Deduct 18 from the shortest and 11 days from the longest cycle. Thus if a woman's cycle varies from 24 to 32 days (best recorded over 12 menstrual cycles)

$$24 - 18 = 6$$
$$32 - 11 = 21$$

The first 5 days of the above example are regarded as 'safe' for a couple to partake in intercourse. After day 21 to the end of the cycle is also 'safe'.

2 Method (2): The woman takes her temperature once a day, first thing in the morning before getting out of bed, eating or drinking. Basal body temperatures rises by about 0.5°C at mid-cycle and remains high until menstruation starts. Coitus should be avoided before and for at least 3 days after ovulation. This method obviously requires determination and persistence and a good thermometer.

3   Method (3): Testing of vaginal secretion. The texture of the vaginal secretion changes from its normal slippery to a sticky nature just before ovulation. Coitus should be avoided at this time.

The safe period is the only method approved by the Roman Catholic Church. However, it requires intelligent and persistent patients with regular cycles. Continence is required over part of the cycle.

Summary of safe period method:

1   Effectiveness – High failure rate: 15–35 pregnancies/100 woman years (wy)

2   Cost – low

3   Acceptability – moderate

4   Availability – universal

5   Contraindications – none

6   Side effects – none

METHODS USED BY THE MALE

Condom or sheath (See Appendix 1)

Most condoms (sheath, protective, French letter etc.) are disposable and are available in different qualities of latex of different thickness and in various colours. A condom should cover the whole penis – 'American Tips' are unreliable. The condom also offers some protection from sexually transmitted infections. For best results the condom should be withdrawn immediately after ejaculation and be used simultaneously with chemical spermicides (cream, jelly, paste, foam or pessary).

1   Effectiveness – 1–15 pregnancies/100 wy; less than 1–5 with spermicide back up

2   Cost – high

3   Acceptability – moderate

4   Availability – excellent

5   Contraindications – none

6 Side effects – rare rubber allergy – special condoms can be obtained

Coitus interruptus

Probably the commonest method worldwide. The man withdraws his penis from the vagina before ejaculation. This is a risky practice and requires discipline and strong motivation. For some couples coitus interruptus may be psychologically harmful and may prevent full sexual enjoyment.

1 Effectiveness – high failure rate: 3–17 pregnancies/100 wy

2 Cost – none

3 Acceptability – moderate

4 Availability – universal

5 Contraindications – in cases of impotence, frigidity, premature ejaculation or intermittent ejaculation

METHODS USED BY THE FEMALE

The diaphragm or 'dutch cap' (See Appendix 1) (Figure 8.1)

These consist of a coiled spring with a dome of soft rubber, and are available in sizes from 50 mm to 105 mm diameter at 5 mm

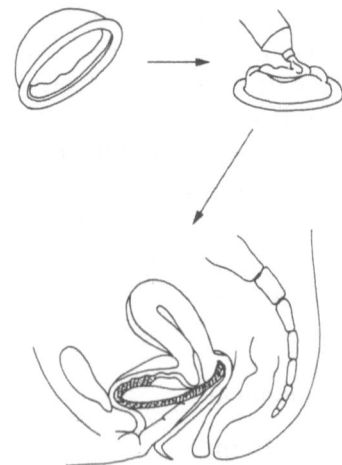

Figure 8.1 *Insertion of vaginal diaphragm*

intervals. The diaphragm should be used with chemical spermicides, and acts as both a mechanical barrier and as a vehicle to convey the chemical spermicide to the cervix. Two 2.5 cm strips of tubed spermicide (cream or jelly) should be applied to the convex side of the diaphragm and smeared over the surface and round the rim. Two further 2.5 cm strips of spermicide should be applied to the concave surface. The diaphragm or cap should not be worn for more than 24 hours without removal for cleaning. It should remain in place for at least 6 hours after the last intercourse. Rechecks are needed at two year intervals; weight gain may change requirements.

1 Effectiveness – with spermicide 1–5 pregnancies/100 wy

2 Cost of spermicides – high

3 Acceptability – poor: lack of spontaneity unless fitted nightly

4 Availability – the correct size of diaphragm must be selected by trained personnel, and the user should re-attend to check the fitting after one week, three months, and then annually. After childbirth, any vaginal operation, weight gain, re-fitting is necessary at the post-natal visit

5 Contraindications:
   a Prolapse – marked degree of cystocoele or urethrocoele
   b Abnormal pelvic shape or deficient perineum. Under such conditions the cervical cap may be used
   c Aesthetic contraindications

6 Side effects – rare rubber allergy

Cervical cap

Cervical caps are made of rubber and are thimble shaped with a raised rim designed to fit closely round the base of the cervix. They are available in four sizes: small, medium, large and extra large. Cap size is selected so that cervix is neatly contained and the rim just touches the vault of the vagina. The smallness of the cervical cap is usually more acceptable than the diaphragm. Whereas the diaphragm should be removed within 24 hours after insertion, the cap can be left in place for up to four days, though the technique of fitting is more difficult to teach and learn. Before it is inserted a 5 cm strip of tubed spermicide should be applied to both sides of the cap

1   Effectiveness – with spermicide 1–5 pregnancies/100 wy

2   Cost with spermicide – high

3   Acceptability – poor

4   Availability – poor

5   Contraindications:
    a   Cervical erosion
    b   Chronic cervicitis
    c   Where the cervix points backwards
    d   Cervical tear extending to the base of the cervix
    e   Patient fitting difficulty

6   Side effects – rare rubber allergy

The vault (Dumas) cap

This is an intravaginal rubber cap that fits across the cervix and adheres by suction to the fornices. It is particularly useful in multiparous women with poor pelvic floor muscle tone and where the cervix is not suitable for a cervical cap. The correct size (from 50 to 75 mm in 5 mm steps) is the smallest that fits evenly into the vaginal vault. Before it is inserted a 5 cm strip of tubed spermicide should be applied to each side of the cap.

1   Effectiveness – with spermicide 1–5 pregnancies/100 wy

2   Cost with spermicide – high

3   Acceptability – poor

4   Availability – poor

5   Contraindications:
    a   Long cervix
    b   Backward pointing cervix
    c   Patient fitting difficulty

Spermicides (See Appendix 2)

Spermicides are chemicals such as the long-chain alcohol

nonoxynol-9 or simpler substances such as boric acid that are
capable of killing spermatozoa. They are available in the form of:

1 Pessaries

2 Foaming preparations

3 Creams and jellies (tubed products)

4 Pastes and sponges

5 Aerosol foams

Note: Whenever a condom, diaphragm, or cervical cap is used a
spermicide should also be applied.

1 Effectiveness – high failure rate: 10–40 pregnancies/100 wy
   when used alone

2 Cost – high

3 Acceptability – poor

4 Availability – good, mainly used as adjuvant to mechanical
   methods

5 Contraindications – occasional allergy

6 Side effects – none

C-film, a small piece of water-soluble plastic material impreg-
nated with spermicide and placed either over the penis or over
the cervix is not reliable. It has a failure rate of about 60 per 100
woman years.

Intrauterine devices (See Appendix 3)

Types of IUD: Most are made of barium-impregnated, flexible
polythene or nylon, in various shapes (see Figure 8.2). Recent
improvements include impregnation with copper ('Copper 7',
'Novagard', Gravigard, Multiload Nova T, Ortho-Gyne T)

Mode of action: This is uncertain. Various theories have been
suggested but the exact mode of action is not known:

1 Tubal peristalsis may be accelerated; this may hinder
   implantation

2 The IUD placed against the endometrium may act as a
   physical barrier to implantation

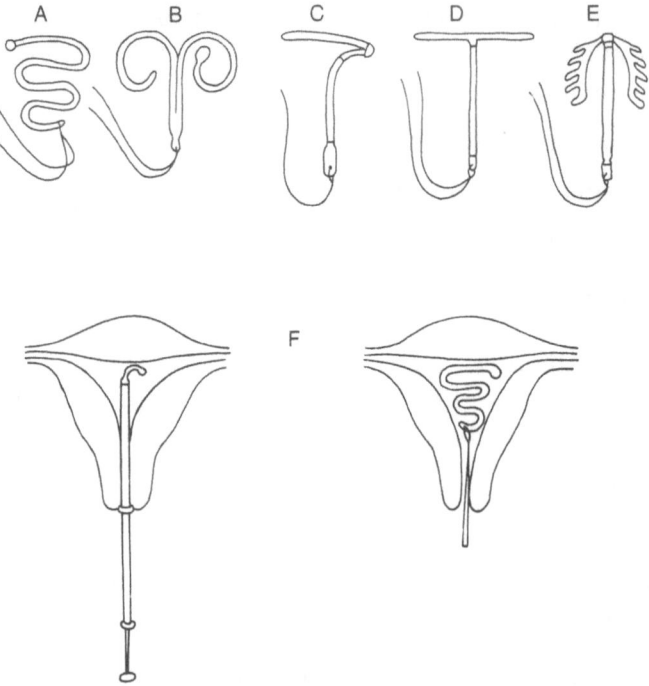

Figure 8.2 *Intrauterine devices A Lippes Loop; B Saf-T Coil; C Copper 7; D Copper T; E Multiload Copper-250; Insertion of the Lippes Loop*

3 The IUD may cause a subclinical abortion of a fertilized embryo

4 The device causes a chronic inflammatory response via a macrophage (round-cell) infiltration of the endometrium, and at possible implantation sites

5 Metallic copper may greatly enhance the effect of the IUD by a direct anti-fertility effect

Indications:

1 Patient choice

2 Other methods unsuitable or unacceptable

3 For the impulsive and casual or where continuous motivation is a problem

4 Underdeveloped countries

Contraindications:

1 Menorrhagia and abnormal uterine bleeding

2 Acute pelvic infection, chronic cervicitis

3 History of pelvic inflammatory disease especially gonococcal

4 Pregnancy

5 History of ectopic pregnancy especially in nulliparous patient

6 Suspected malignancy of the genital tract

7 Congenital abnormalities of the uterus, e.g. bicornuate uterus

8 Fibroids – especially if these impinge on the uterine cavity. Symptomless sub-serous or intramural fibroids are not contraindications

9 Need for 100% contraception (unless can be 'backed' by termination of pregnancy)

10 Known allergy to copper if a copper IUD is to be inserted

11 Nulliparity

12 Caesarian section

Timing of insertion:

1 Interval when not pregnant. Preferably during or just after a period

2 Postpartum – routine 8–12 weeks (12 weeks for Caesarian section)

3 Post-abortion or after termination of pregnancy – stat or 6 weeks

Method of insertion:

1 Insertion should be preceded by counselling and advice so that the patient clearly understands the advantages and disadvantages of the method

2 History: exclude contraindications

3 Examination:
   a Pelvic examination excludes uterine abnormalities and ascertains position and size and mobility of the uterus
   b Cervical smear for exfoliative cytology

4 Insertion (follow meticulously the manufacturers instructions). Some general comments:
   a Routine bimanual examination
   b Introducing a Cusco vaginal speculum exposes the cervix – this is then cleaned with antiseptic solution
   c The anterior lip of the cervix may be grasped with tenacular forceps. A sterile uterine sound is introduced to assess the length and direction of the uterine cavity
   d If the IUD has an introducer the sterile IUD is loaded into the introducer. The introducer is then inserted into the uterus, to reach the fundus
   e The IUD is freed from the introducer and left in the fundus. The thread attached to the IUD is cut, about 3 cm from the cervix: this enables the patient to check that the IUD is in place, and facilitates its removal later

Note: Copper dispensing IUDs need to be replaced about every 2–5 years. See individual manufacturers recommendations

Insertion problems:
1 Colic – sometimes may necessitate the removal of the device

2 Syncope and shock – especially in nulliparous and nervous patients (vaso-vagal reflex)

3 Tetany

4 Difficulty with insertion

5 Perforation

Post insertion problems:
1 Bleeding:
   a Menstrual
   b Intermenstrual
2 Pain:
   a Colic

    b Dysmenorrhoea

3 Vaginal discharge is a common side effect

4 Pregnancy (one in twenty pregnancies are ectopic)

5 Pelvic inflammatory disease – special problem in the young nulliparous women

6 Perforation – probably always at insertion – about 1/1000 (Lippes C & D commonest)

7 Expulsion – of all spontaneous expulsions, 50% occur in first 3 months after insertion

8 Male dyspareunia – if threads are cut too short or if the IUD is protruding through the cervical os

9 Encrustation – deposition of calcium salts – usually confined to devices containing barium or bismuth salts

Follow-up visits:
1 First check visit – 3 months after insertion – preferably at the clinic where the device was fitted
2 Annual check visit – by any doctor

Continuation rate: This depends on:
1 Device
    a Size, shape, texture and surface contact
    b Simple to insert
    c Easy to retract

2 Doctor:
    a Skill, experience, attitude
    b Training

3 Patient:
    a Motivation and attitude
    b Age, parity

Missing threads – management:
1 Probe cervical canal with long Spencer Wells forceps

2 Sound uterine cavity

3 Probe cavity with hook (Fish!)

4 Ultrasound investigation

5 X-ray examination

6 Laparoscopy

7 Laparotomy

In about 90% of cases with missing threads the IUD is still in the uterus

Summary of IUDs:
1 Effectiveness – around three pregnancies per 100 woman years. If pregnancy occurs the IUD should be removed if this can be done easily. Should the woman elect not to terminate the pregnancy, there is a 30–40% chance of a spontaneous abortion if the coil is *in situ*. The rate of ectopic pregnancy is also increased from about 0.4% to about 0.8%

2 Cost – high initially

3 Acceptability:
  a They are effective immediately after insertion
  b There is a return to normal fertility after removal
  c Does not irritate either partner; especially useful in poorly motivated couples
  d Can be left in place until the menopause, or until a further pregnancy is planned

4 Availability – poor: must be inserted by a trained operative. Almost always easy to insert

5 Side effects – minor side effects:
  a Intermittent bleeding and abdominal cramps – usually occur within the week following insertion (around 10% of cases in first year)
  b Increased, heavier periods
  c Expulsion may occur in 10% of cases

6 Major side effects:
  a Uterine perforation
  b Reactivation of pelvic infection. Rare unless there was a previous history of pelvic infection

HORMONE CONTRACEPTION (See Appendix 4)

Since about the 1930s it has been known that the sex hormone steroids oestrogen and progesterone can inhibit ovulation in women. The last five years or so have seen great developments in this area and there has been a reduction in dosage without a change in contraceptive efficiency. At present about 50 million women are using oral contraceptives. In Britain this includes about 2.5 million women; about one in five of all married couples.
Types:

1 Combined oestrogen and progestagen products commonly referred to as 'The Pill'. Sequential oral contraceptive pills are little used today

2 Progestagen only products commonly referred to as the 'Mini-Pill' or 'POPS'

3 The triphasic pill, a new development, has a variable dose of oestrogen and progestagen that mimic the normal endocrine changes over the menstrual cycle. The pill is principally oestrogenic in the first part and principally progestagenic in the second half of the course. Triphasic pills contain the lowest total combined monthly dosage compatible with effective contraception

4 Injectable hormonal contraception – intramuscular injection every three months of progestagen alone or combination of oestrogen and progestagen

Oral contraceptive agents

Beneficial effects of oral contraceptives:

1 Majority of women experience a sense of well-being

2 Prevention of pregnancy – oral contraceptives are virtually 100% effective and acceptable

3 Pelvic inflammatory disease – reduction in incidence

4 Menstrual cycle – decreased menses, regulation of irregular cycles, decrease in incidence of premenstrual tension, dysmenorrhoea and endometriosis, relief of mid-cycle (ovulation) pain

5 Neoplasia – protective effect against benign breast tumours and benign ovarian cysts. Decreased risk of ovarian and endometrial cancer

6 Rheumatoid arthritis – reduction in incidence

Mechanism of action:
1 Progestagens:
   a Depress the mid-cycle surge in follicle stimulating hormone (FSH) and luteinizing hormone (LH) as well as the baseline levels. This prevents ovulation. However this may not always occur with low doses of progestagens, but there is always some reduction in the peak levels of the gonadotrophins
   b Alter the quality of cervical mucus, making it thick and viscous and less penetrable to spermatozoa
   c Produce endometrial changes and altered motility of uterus and oviducts

2 Oestrogens:
   a If given from the first days of menstruation, both FSH and LH are depressed
   b If given beginning from day 7 or 8 of the cycle, they cause a decline in FSH levels and raised LH levels. Hence there is no ovulation, but the elevated LH does luteinize the follicle. Hence a new cycle will commence, and ovulation occurs about two weeks later.

Thus when oral contraceptives are taken, there is no true menstrual cycle but hormone withdrawal bleeds occur during the pill-free intervals (see below).

Combined oral contraceptives

All combined oral contraceptives recommended by the Family Planning Association consist of one of two oestrogens:
1 Ethinyloestradiol (50, 30 or 20 μg)

2 Ethinyloestradiol 3-methyl-ether: Mestranol (50 μg) and one of five progestagens:
   a Norethisterone

   b Norethisterone acetate

 c Desogestrel

 d Ethynodiol diacetate

 e Norgestrel

Particular indications for using oral contraceptives:
1 General:
 a Young fit single woman
 b Young married couples not yet planning to have children
 c After termination of pregnancy
 d Mothers already with children, not desiring more
 e Unwillingness to use other methods of contraception

2 Specific:
 a Parents having inheritable genetic abnormalities that they do not wish to transmit to their offspring
 b Non-contraceptive uses:
  i Irregular menstrual cycles
  ii Dysmenorrhoea
  iii Menorrhagia
  iv Endometriosis

Prescribing the pill

Before prescribing the pill, attention should be paid to the following:
1 History: Enquire in particular for:
 a Past thromboembolic disease
 b Liver disease
 c Diabetes mellitus
 d Malignant disease particularly of the breast or genital tract
 e Menstrual history, in particular of amenorrhoea
 f Family history of coronary artery disease or of hypertension
 g Age and smoking habits

2 General examination
 a Body weight
 b Blood pressure

    c Breast examination

    d Varicose veins

    e Test urine for sugar and protein

3 Pelvic examination

    a Vulva

    b Vagina

    c Cervix

    d Uterus – size, position, mobility

    e Adnexae

4 Speculum examination

    a Condition of cervix

    b Nature of discharge

    c Cervical smear cytology

5 Choice of contraceptive: This will vary with the clinician. In general, begin with the lowest dose of oestrogen

6 Regime: Varies in particular circumstances:

    a First course beginning on first day of menstrual period. This means the first period will be shorter than normal; after this the regime goes on to that described in **b** or

    b First course begun on fifth day of period. Pill then taken for 21 days with a 7 day intermission, when menstruation should occur. Extra precautions are needed for first 14 days or

    c Pill taken for 84 days, followed by 7 pill-free days

    d Special circumstances:

        i Pill can be started from 2–4 weeks post partum, if patient not breast feeding

        ii After termination (or miscarriage) pill may be started the day after

7 Urge strict daily pill taking; if one day is forgotten 'double up' the next day, preferably within 12 hours of missed pill

8 If a change of pill is required to a lower dose oral contraceptive or a progesterone-only pill extra contraceptive precautions should be used for the first month

9 Postponing a period is effected by taking two courses of the pill consecutively

Contraindications to oral contraceptive agents:
1 Absolute contraindications:
  a History of thromboembolic disease
  b Severe heart disease
  c Cerebrovascular accident
  d Blood dyscrasia, in particular sickle cell disease
  e Undiagnosed abnormal vaginal bleeding
  f Liver disease – Dubin–Johnson or Rotor syndromes; recent infective hepatitis or glandular fever
  g Nulliparous women with poor ovarian function or with recognized pituitary dysfunction
  h Malignant disease of breast or genital tract
  i Pregnancy

2 Relative contraindications:
  a Varicose veins
  b Porphyria
  c Migraine
  d Epilepsy
  e Contact lenses
  f Fibroids
  g Diabetes mellitus
  h Gall bladder disease
  i Renal disease
  j Mild hypertension or a history of hypertensive disease of pregnancy
  k Herpes of pregnancy
  l Over age 40, or smoking and age over 35 years
  m Amenorrhoea or hypomenorrhoea in the young
  n Surgery – oral contraceptives should be discontinued one month before surgery

Systemic effects of oral contraceptive agents:
1 Increase in blood pressure. In most patients this is a small and reversible effect:
  a Approximately 5% of users over a 5 year period showed a small increase in systolic pressure. This may relate

to an elevated angiotensin level and also slightly raised aldosterone levels

b However, the majority of these patients had slightly elevated blood pressures before beginning: an increase in blood pressure in normotensive women was less common and reversed once medication was stopped

2 Women taking the pill who smoke and/or are hypertensive show a substantial excess death rate from myocardial infarction.

    a A UK study has shown that, in the 30–35 year age group, the total incidence of myocardial infarction was 11· per 100 000, whereas age-matched controls had a risk of 4 per 100 000

    b The risk increases markedly with:
       i Increasing age
      ii Dose of oestrogen in the pill
     iii Abnormal carbohydrate metabolism
     iv Cigarette smoking
      v Obesity
     vi Family history of cardiovascular disease

    c It is thought that the presence of:
       i One risk factor increases by a factor of 4× the base line risk
      ii Two risk factors increases by a factor of 10× the base line risk
     iii Three risk factors increases by a factor of 80× the base line risk

    For example, in the same study, for women in the 40–44 year age group, the incidence of myocardial infarction was 112 per 100 000 as compared to 22 per 100 000 controls

3 Increase in concentration of blood lipids: raised levels of triglyceride, cholesterol and lipoproteins, due to increased production by the liver

4 Impairment of glucose tolerance – 'steroid diabetes'. Diabetes could be unmasked and existing diabetes could be made more difficult to control. This may be a progestagenic effect

5 Alteration of clotting factors – increased levels of clotting factors and increased platelet aggregation – this increases the incidence of thromboembolism. The oestrogen component is the main associated factor. Thromboembolic problems represent the most serious risks of the oral contraceptive agent:

    a Retrospective studies suggested that the risk of death from thromboembolism was 1.30 per 100 000 women (cf. 0.2: 100 000 in controls not on the pill)

    b Similar data for the United States give an incidence of between 4 and 6 per 100 000 mortality due to thromboembolic phenomena. However, these absolute figures are in fact small. Furthermore

    c A prospective study (in 1972) involving 80 000 women failed to demonstrate an increase in the incidence of thromboembolic phenomena

6 Effects on endocrine function:

    a Thyroid function – increase in levels of blood protein bound iodine (PBI) and of thyroid binding globulin (TBG)

    b Adrenocortical function – increase in corticosteroid binding globulin (transcortin)

    c Anterior pituitary – reduced sensitivity to gonadotrophin stimulation

7 Changes in liver function:

    a Increased levels of the liver enzymes aspartate transaminase (AST) and alanine transaminase (ALT)

    b There is an increase in the levels of most plasma proteins (caeruloplasmin, transferrin, and transcortin) and a decrease in serum albumin

    c Increase in incidence of cholestatic jaundice and gallstone formation

8 Effects on skin: Increase in chloasma (melasma), acne (with low oestrogen dosage), photosensitivity, telangiectasis and diffuse alopecia

9 Effects on central nervous system:

    a Increased incidence of nausea and vomiting as well as migraine and headaches

   b Alteration in tryptophan metabolism – reduction in
         brain and sleep disturbances

10 Menstrual problems:
   a Monthly bleeding (withdrawal bleeding) is reduced
         and this is related to the anti-oestrogenicity of the pro-
         gestagen used
   b Amenorrhoea during treatment with the pill may
         occur. If it occurs for the first time, the patient should
         continue into the next cycle. However if it continues
         for more than 3 cycles:
      i Pregnancy should be suspected
      ii Treatment should be terminated, or a different
            pill used
   c Prolonged amenorrhoea after stopping the pill should
         be investigated – it is sometimes associated with
         hyperprolactinaemia
   d Breakthrough bleeding ('spotting'), if it occurs beyond
         the first cycle a change of pill with a higher oestrogen
         content may be needed

11 The pill: *Lancet* reports and cancer
   Two papers were published in the *Lancet* in October 1983
   which suggested a link between breast and cervical cancer
   and the oral contraception pill. The Family Planning
   Association believes the following low dose pills are pre-
   ferred for women of all ages taking oral contraception

Combined pills	Biphasic pills and triphasic pills	Progestagen only pills
Neocon 1/35	BiNovum	Micronor
Norimin	Logynon	Noriday
Brevinor	Logynon ED	Microval
Ovysmen	Trinordiol	Norgeston
Loestrin 20		Neogest
Ovranette		Femulen
Microgynon 30		
Marvelon		

This list is based on present evidence and it is likely that
there will be additional information in the near future.

Side effects of oral contraceptive agents:
Major side effects:

1 Deep vein thrombosis

2 Pulmonary embolism

3 Cerebral thrombosis

4 Cerebral haemorrhage

5 Hypertension

6 Myocardial infarction

7 Severe depression

8 Severe migraine/headache

9 Gall bladder disease

10 Jaundice

11 Exacerbation of diabetes

12 Exacerbation of epilepsy

Minor side effects:

1 Persisting nausea – usually noticed less if tablet taken last thing at night

2 Breast pains and tenderness – usually disappear in subsequent cycles

3 Breakthrough bleeding

4 Weight gain of about 1 kg

5 Loss of libido

6 Anxiety and to a lesser degree depression

7 Headaches

8 Failure of vaginal lubrication

9 Recurrent vaginal candidiasis

10 Cervical erosion

11 Facial hirsutism

12 Ankle oedema

13 Increase in acne

14 Abdominal bloating

15 Chloasma

16 Irritation from contact lenses

17 Galactorrhoea

Drug interactions with oral contraceptives:
1 Rifampicin, phenytoin and phenobarbitone – more needed to maintain same therapeutic effect since oral contraceptives increase drug metabolism in the liver

2 Ampicillin and tetracyclines reduce the absorption of oestrogen and progestagens and may cause contraceptive failure

3 Oestrogens affect glucose tolerance and the dose of hypoglycaemic agents (insulin or oral preparations) may need to be increased

4 Corticosteroids act in direct competition in the metabolism of oestrogens and progestogens

Follow-up visits

Routine pill checks should be made every 3 months for the first year and every 6 months thereafter.
Enquire for:
1 Last menstrual period

2 Cycle control – any spotting or breakthrough bleeding

3 Withdrawal bleed or amenorrhoea

4 Any side effects – headaches, nausea, breast discomfort, vaginal discharge

5 Ask for:
   a Cough
   b Breathlessness
   c Pain in the chest
   d Pain in calf and legs

6 Examine
    a Body weight – gain of over 2 kg is undesirable
    b Blood pressure

Continuation depends upon:

1 Development of complications while on the pill

2 Feelings and anxieties expressed by the patient

3 Risk factors due to, and side effects of, the pill

When stopping the pill, unless there is an important reason, the cycle should be completed. If planning a pregnancy, it is usually recommended that one normal menstrual period be allowed before trying to conceive. Most women will experience a delay of up to 20 days in ovulation: this must be taken into account when estimating the expected date of delivery

Summary of combined oral contraceptives

1 Effectiveness: Good – 0.025–1 pregnancy/100 wy

2 Cost: Fairly low

3 Acceptability: Fairly good

4 Availability: Fair

5 Contraindications: Multiple – see above

6 Side effects: Multiple – see above

Progesterone-only pill or injections (See Appendix 4)

Many of the unwanted side effects and thromboembolic phenomena are attributed to the oestrogenic component of oral contraceptives. Unlike combined pills progesterone-only pills (POPs) are taken every day throughout the cycle including during periods.
Mechanism of action:

1 Cervical mucus becomes thicker and acts as a barrier to sperm

2 Suppression of endometrium

3   Altered mobility of uterus and tubes

4   Ovulation is inhibited in only about 40% of cycles

Examples of progesterone-only pills currently available (see Appendix 1):

Femulen: 500 µg ethynodiol diacetate
Micronor: 350 µg norethisterone
Noriday: 350 µg norethisterone
Neogest: 75 µg DL-norgestrel
Microval: 30 µg levonorgestrel
Norgeston: 30 µg levonorgestrel

Systemic effects of progesterone-only pills:
1   POPs are able to reverse the accelerated clotting patterns in women who have been on combined preparations. No change in clotting factors has been shown. Some alterations do occur in the platelet aggregation factor but this is consistently lower than with the combined pill

2   No significant effect on:
    a   Glucose tolerance
    b   Serum lipid
    c   Thyroid function
    d   Liver function
    e   Adrenal function

Side effects
1   Irregular bleeding

2   Amenorrhoea

3   Weight gain, loss of libido, breast tenderness, headache, and acne are reported but occur with less frequency than with the combined pill

Injectable hormonal contraception depot progesterones

In many parts of the world a progestagenic drug, medroxyprogesterone acetate (Depo-Provera), given by intramuscular injection is widely used for contraception. Recently doubts have been raised on its future and at present it is only available for short-term use.

Indications:

1 For the consorts of men undergoing vasectomy for protection until vasectomy becomes effective

2 For women who are being immunized against rubella to prevent pregnancy during the active phase of the virus

Administration:

Single intramuscular injection of 150 mg Depo-Provera provides protection against pregnancy for about 3 months. Best given at the beginning of the menstrual cycle. Extra precautions should be used for the first 14 days after the injection.

Contraindications – similar to contraindications to oral contraceptives

Side effects – same as progesterone-only pill

Summary of POPs and depot progesterone injections

1 Effectiveness: 1–5 pregnancies/100 wy

2 Cost: Fairly low

3 Acceptability: Poor to moderate

4 Availability:
   a POPs: Fair
   b Injection: Poor

5 Contraindications: Thromboembolic episodes (see oral contraceptive section)

6 Side effects: Irregular bleeding, higher failure rate

Conclusion:

Hormonal contraceptives are the most efficient and reliable method of temporary contraception. The risks relating to the combined oral contraceptive pill are small when considered in perspective. The risk of 0.3–3.0 deaths from oral contraception per 100 000 woman years is low compared to:

1 20–23 deaths per 100 000 pregnant women in pregnancy

2   3.2 deaths per 100 000 elective abortions

3   25–30 deaths from road accidents per 100 000 person-years

POSTCOITAL CONTRACEPTION

Indications:

1   'Emergencies': women who have had coitus using no contraception at the time of ovulation

2   Women who have had intercourse using a method of contraception which has failed, e.g. a burst sheath

3   Women who have coitus infrequently and who do not wish to take medication every day

Preparations:

1   Eugynon 50 or Ovran – four pills given in two doses up to 72 hours after intercourse prevents implantation of the fertilized ovum

2   Insertion of a copper intrauterine contraceptive device (Copper 7 or Copper 250) up to 5 days after intercourse

	Postcoital pill	*Postcoital intrauterine device*
Method:	Eugynon 50 or Ovran Two pills STAT Two pills 12 hours later	Any IUD inserted postcoitally
Timing after intercourse	Up to 72 hours	Up to 5 days
Efficacy	99%	100%
Side-effects	Nausea Vomiting	Pain and bleeding Risk of infection
	– Potential teratogenic effect –	
Contraindications	History of thrombosis Ectopic pregnancy	Ectopic pregnancy

STERILIZATION

In the United Kingdom anyone of 16 or more years of age can consent to a sterilization procedure provided the operation is performed for a good reason. Where there is a legally married cohabiting couple, the written consent of the spouse should be obtained before operation.

Medical indications for sterilization occur when pregnancy would become a hazard to the woman's life or future health. However, most who opt for sterilization are healthy women who do not want any more children. About 20% of all married couples in the USA have had one partner sterilized.

Initial management

Both partners involved should be seen. A careful history should be taken, and the couple should be informed that the procedures are essentially irreversible. Although microsurgical techniques may give good results in re-uniting the oviducts or vas deferens, they are not universally available and success cannot be guaranteed.

Which partner should be sterilized:

1 Emotional acceptance

2 Age

3 Least healthy?

4 Genetic – dominant genes

Causes of regret poststerilization:

1 Sterilization for medical reasons

2 Marital disharmony

3 Youth

4 Following termination of pregnancy

5 Immediately postpartum

6 During depression

Background counselling and discussion for sterilization/
vasectomy:

1   Age of husband and wife and occupations

2   Number of years married

3   Number and ages of children

4   Abortions and any unplanned pregnancies

5   Husband's general health

6   Wife's health
  a   General
  b   Obstetrical
  c   Gynaecological

7   Children's health

8   Previous contraception, coital-frequency, satisfaction, difficulties

9   Couple's reason for sterilization

10 Consideration of loss of child, loss of spouse or marriage breakdown

Vasectomy

Definition: The surgical removal of a section of the vas deferens;
the remaining ends are turned back on themselves and tied.
Details of vasectomy:

1   Operation must be regarded as irreversible

2   Consent of GP usually essential

3   Details of operation
  a   Anatomy
  b   Technique
  c   Local anaesthetic
  d   Removal of vas and suturing
  e   Absorbable skin sutures
  f   Advisable to take ... days off work (depends on nature of job)

4   Postoperative visit after one week

5 Effect of the operation
  a Testes and other accessory glands are unaffected by the operation
  b Spermatozoa are absorbed in the proximal part of the vas. Seminal fluid still produced

6 Post-operation fertility exists for up to 3 months
  a Continued intercourse or ejaculation needed to 'remove' residual sperms from tubes distal to operation site
  b Other method of contraception must be used over this period

7 Tests for sterility:
  a Semen specimens at 12 and 16 weeks postoperatively
  b If longer – test semen until two consecutive specimens taken at monthly intervals show absolutely no sperm

Advantages of vasectomy:
1 Simple technique – this can be done on an outpatient basis under local anaesthesia

2 Little morbidity

3 Very effective

Disadvantages of vasectomy:
1 Male and female attitudes often make it unacceptable

2 Reversible in only 20–40% of cases

3 Sterilization is not effective for at least three months and other methods of contraception should be used over this time

Contraindications:
1 Instability of the marriage

2 A psychiatric history

3 When the wife may need hysterectomy in the near future

4 Medical reasons:
  a Hernia or repair

b Undescended testicles
c Other genital abnormality
d Local anaesthetic reaction

Complications:
1 Haematoma formation and subsequent infection

2 Wrong structure may be tied

3 Wound infection

4 Emotional problems

5 Increases in serum:

a Uric acid concentration
b Testosterone concentration
c Sperm agglutinating antibodies

6 Recanalization of the vas – very small risk

7 Long-term complications may rarely occur and include
a Secondary infection
b Orchitis
c Epididymitis
d Spermatic granuloma
e Sinus formation

Tubal sterilization

10–20% of women in the UK are sterilized by age 40
Points to discuss:
1 Is the request genuine or half-hearted? What does her husband think of her request? His consent is preferable but it is not legally necessary

2 Does she understand the finality of the procedure? Explain the effect of the operation on menstruation

3 Does she understand the nature of the procedure to be employed? Explain that about 1 in 1000 laparoscopic sterilizations fail
a Laparoscopic sterilization
Method: A laparoscope is inserted into the peritoneal

cavity under anaesthesia and insufflation of the abdominal cavity with CO_2. The Fallopian tube is then located, and grasped. It is then cauterized, or alternatively and more commonly clips or rings are applied
Disadvantages:
- i Requires general anaesthetic, and an incision into the peritoneal cavity. However this procedure involves less disturbance to the patient than does formal tubal ligation
- ii Not reversible; however there may be a greater chance of reversal (if this is required) if clips rather than cautery are used
- iii Possibility of damage to other organs
- iv Infection, haemorrhage
- v Overdistension of the abdomen with insufflated gas can cause cardiac arrhythmias

b Tubal ligation/excision:
Method: Abdominal incision. A loop of tube is tied, and excised, or its end is buried in the broad ligament
Disadvantages:
- i Requires anaesthesia
- ii Peritoneal cavity is opened; there is an associated morbidity and mortality, however small
- iii Not generally reversible

c Vaginal sterilization:
Method: Incision through posterior vaginal fornix – simple operation – commonly performed in India
Disadvantages:
- i Substantial morbidity as a result of infection
- ii Dyspareunia

Failure rate: About 1–2 per 1000 women become pregnant after tubal ligation or laparoscopic sterilization
Causes of failure:
1 Many failures show pre-existing pregnancy

2 Ring or clip placed on round ligament or appendix

Reversal: The results for reversal of female sterilization improve year by year. It varies for the initial type of operation performed:

Diathermy	10% success rate
Tubal ligation	40% success rate
Rings	50% success rate
Clips	70% success rate

It is thought that about 5% of women who were previously sterilized want reversal and another 5% enquire about reversal.

This is because of:

1 Change of partner

2 Psychiatric reasons – 'Not felt right'

Complications of sterilization: Side effects are rare

1 Menstrual problems – menorrhagia – but trials are equivocal

2 Loss of libido

3 Recanalization

4 Ectopic pregnancy

Hysterectomy

In the USA hysterectomy is considered as an alternative to laparoscopic sterilization. Elective hysterectomy is the commonest surgical operation performed in the States.

Comparison of laparoscopic sterilization and hysterectomy

		Sterilization	Hysterectomy
1	Admission	Day case	Minimum 5 days
2	Mortality	1/16 000	1/1600
3	Morbidity	4%	30%
4	Postoperative malaise	+	+++
5	Future hysterectomy	15%	0
6	Carcinoma uterus	1/30	0
7	Profit and cost	+	+++

APPENDIX 1

Barrier devices

Diaphragms and caps for use with spermicide:
All diaphragms and caps available are made of rubber

Product	Description
Diaphragms	
Durex	Flat spring diaphragm
Ortho	Coil spring diaphragm
Caps	
Dumas	Vault cap
Prentif Cavity Rim	Cervical cap
Vimule Cap	Vimule cap

Condoms:
Non-lubricated condoms for use with spermicide, e.g. Durex Allergy, etc. Lubricated for use with spermicide, e.g. Atlas, Durex Black Shadow, etc. Durex Nu-form Extra Safe sheath incorporates a spermicidal lubricant.

APPENDIX 2

Spermicidal contraceptives

The FPA recommends that spermicides should only be used in conjunction with a barrier method of contraception
1 Foams – supplied in aerosol containers, with or without an applicator:
  a Delfen Foam
  b Emko Foam

2 Creams – supplied in metal tubes:
  a Duracreme
  b Orthocreme

3 Jellies – supplied in metal tubes:
  a Duragel
  b Ortho-Gynol Gel
  c Staycept Jelly

4 Pessaries – for use with condoms, diaphragms and caps:
   a Orthoforms
   b Staycept Pessaries
   c Double Check

5 Pessaries – for use with CONDOMS ONLY – not to be used with diaphragms and caps:
   a Genexol
   b Rendells
NB Genexol and Rendells should not be used with diaphragms or caps, as on prolonged contact they damage rubber

APPENDIX 3

Intrauterine devices

Inert intrauterine devices:

Product	Recommendation	Description
Lippes Loop (Sizes B, C and D)	For use in multiparous women	White polyethylene with radio-opaque substance added Size B – Black nylon thread Size C – Yellow nylon thread Size D – White nylon thread

Bio-active intrauterine devices:

Product	Recommendation
Copper-7 (Gravigard) (Mini-Gravigard)	For use in multiparous and nulliparous women
Copper-T (Ortho Gyne-T)	For use in multiparous and nulliparous women
Multiload Cu250	For use in multiparous and nulliparous women
Multiload short-stem	For use in multiparous and nulliparous women
Novagard	For use in multiparous and nulliparous women
Nova T	For use in multiparous and nulliparous women

APPENDIX 4

Oral contraceptives

The combined oestrogen–progestagen pill is the most effective method of contraception with a failure rate of about 0.1 per 100 woman-years. The progestagen-only pill has a failure rate of about 2–3 per 100 woman years.

Low oestrogen dose preparations:

	Oestrogen	*Progestagen*
Marvelon	0.03 mg ethinyloestradiol	0.15 mg desogestrel
Conova 30	0.03 mg ethinyloestradiol	2.00 mg ethynodiol diacetate
Microgynon 30	0.03 mg ethinyloestradiol	0.15 mg levonorgestrel
Ovranette	0.03 mg ethinyloestradiol	0.15 mg levonorgestrel
Eugynon 30	0.03 mg ethinyloestradiol	0.25 mg levonorgestrel
Ovran 30	0.03 mg ethinyloestradiol	0.25 mg levonorgestrel
Brevinor	0.035 mg ethinyloestradiol	0.50 mg norethisterone
Ovysmen	0.035 mg ethinyloestradiol	0.50 mg norethisterone
Neocon 1/35	0.035 mg ethinyloestradiol	1.00 mg norethisterone
Norimin	0.035 mg ethinyloestradiol	1.00 mg norethisterone
Loestrin 20	0.020 mg ethinyloestradiol	1.00 mg norethisterone
Loestrin 30	0.030 mg ethinyloestradiol	1.50 mg norethisterone

Biphasic and triphasic oral contraceptive: constant dose of oestrogen:
Progestagen dose varies from low at the beginning of the cycle and increases around time of ovulation

	Ethinyloestradiol	*Norethisterone*
BiNovum	0.035 mg	0.5/1.0 mg
TriNovum	0.035 mg	0.5/0.75/1.00 mg

Triphasic oral contraceptive: low oestrogen, low progestagen combination:

	Ethinyloestradiol	*Levonorgestrel*
Logynon	0.03/0.04/0.03 mg	0.05/0.075/0.125 mg
Trinordiol	0.03/0.04/0.03 mg	0.05/0.075/0.125 mg
Logynon ED	0.03/0.04/0.03 mg	0.05/0.075/0.125 mg

Medium oestrogen dose preparation:

	Oestrogen	Progestagen
Ovulen 50	0.05 mg ethinyloestradiol	1.00 mg ethynodiol diacetate
Eugynon 50	0.05 mg ethinyloestradiol	0.50 mg norgestrel
Ovran	0.05 mg ethinyloestradiol	0.25 mg levonorgestrel
Minilyn	0.05 mg ethinyloestradiol	2.5 mg lynoestrenol
Orlest 21	0.05 mg ethinyloestradiol	1.00 mg norethisterone acetate
Minovlar	0.05 mg ethinyloestradiol	1.00 mg norethisterone acetate
Minovlar ED	0.05 mg ethinyloestradiol	1.00 mg norethisterone acetate
Norlestrin	0.05 mg ethinyloestradiol	2.50 mg norethisterone acetate
Gynovlar 21	0.05 mg ethinyloestradiol	3.00 mg norethisterone acetate
Anovlar 21	0.05 mg ethinyloestradiol	4.00 mg norethisterone acetate
Ortho-Novin 1/50	0.05 mg mestranol	1.00 mg norethisterone
Norinyl 1	0.05 mg mestranol	1.00 mg norethisterone

Progestagen-only oral contraceptive:

Femulen	—	0.50 mg ethynodiol diacetate
Microval	—	0.03 mg levonorgestrel
Norgeston	—	0.03 mg levonorgestrel
Neogest	—	0.075 mg norgestrel
Micronor	—	0.35 mg norethisterone
Noriday	—	0.35 mg norethisterone

9

Gynaecological Procedures and Operations

GENERAL SURGICAL MANAGEMENT

Preoperative investigation

1 General:
   a Past medical and surgical history including general
      health
   b Physical examination in particular cardiovascular and
      respiratory systems
   c Urine testing for sugar and albumin

2 Cardiorespiratory system: In the case of any major surgery,
   old patients or any history of chronic disease then, the following to be considered:
   a Chest X-ray
   b Electrocardiogram
   c Physician's opinion
   d Correction of any anaemia present

3 Kidneys/body fluids: In the case of hypertension, history of
   chronic urinary tract infection, the following to be considered:
   a Blood creatinine level
   b Intravenous pyelogram
   c Renal function tests: Creatinine clearance
Any urinary tract infection to be treated before surgery

Preoperative management

1 Mental: Patient and spouse should understand the nature and magnitude of the operation. Both should be aware of the consequences of surgery

2 Bowels: Lower bowel should be empty before all gynaecological operations. Where necessary, laxative may be given 1–2 days before the operation

3 Skin preparation:
   a Skin to be carefully shaved and thoroughly washed
   b Where vagina is to be opened, vaginal wall should be painted with an antiseptic
   c For transvaginal operations, vagina and vulva should be swabbed by hibitane

4 Preoperative physiotherapy: Preoperative training in deep breathing

5 Emptying bladder: Catheterize once patient anaesthetized

Postoperative management

1 Management of pain:
   a Morphine (10 mg) or morphine analogues intramuscularly on recovery of consciousness
   b Change to oral analgesia after first 24 hours

2 Micturition:
   a Catheterize after 20 hours if patient has not passed urine or earlier if patient is uncomfortable
   b After repair operations on vagina it may be necessary to insert a self-retaining catheter under full aseptic precautions and removing this after 5 days

3 Gastrointestinal tract:
   a Use of glycerine suppositories to facilitate defaecation
   b Quantities of fluid and food as tolerated by patient

4 Mobilization:
   a Active leg exercises after 24 hours
   b Deep breathing, to prevent pulmonary complications

c Early mobilization

Postoperative complications

Early complications: 0 to 12 hours postoperatively
1 Pain (see above)

2 Shock:
   a Haemorrhagic shock. A dimunition in blood volume produces a fall in venous return and a fall in cardiac output
   b Cardiogenic shock. Failure of efficient cardiac muscular contraction resulting in a fall of cardiac output. Hypotension results
   c Septic (endotoxic) shock. Due to production of endotoxins from Gram-negative bacteria

3 Management of shock:
   a Cardiorespiratory support
   b Prompt restoration of blood volume. This should be with blood or with plasma or plasma substitutes
   c Intravenous steroids: intravenously 100 mg hydrocortisone or hydrocortisone orally up to 500 mg 6 hourly per kg bodyweight

4 Haemorrhage:
   a Immediate or primary haemorrhage: results from inadequate operative haemostasis
   b Reactionary haemorrhage: results from recovering blood pressure causing bleeding from unligated vessels not fully thrombosed
   c Secondary haemorrhage: occurs around the 10th postoperative day

5 Management of haemorrhage:
   a Stop blood loss, by ligation of any obvious bleeding vessels and gauze packing to any generalized ooze
   b Transfuse if any marked loss of blood

6 Vomiting (very common but usually subsides within 12–24 hours):
   a Intramuscular anti-emetic e.g. metoclopramide

b Replace fluid and electrolyte loss intravenously
c If persistent vomiting occurs beyond 24 hours suspect:
    i Acute dilatation of the stomach
    ii Paralytic ileus or
    iii Intestinal obstruction

Late complications: beyond 12 hours postoperatively:
1 Local complications:
    a Pain
    b Wound problems
        i Bleeding
        ii Dehiscence
        iii Haematoma
        iv Infection (cellulitis or abscess)

2 General complications:
    a Thromboembolic problems:
        i Thrombophlebitis: Local impairment of blood flow due to inflammation of a vein. Managed symptomatically
        ii Deep vein thrombosis: Aseptic thrombosis usually in a deep calf or thigh vein. Predisposing factors include blood stasis, obesity, immobilization, dehydration, anaemia, injury to vein walls and blood changes. Managed by anticoagulation, beginning with intravenous heparin, which is then replaced by therapy with an oral anticoagulant such as warfarin
        iii Pulmonary embolism: Embolization of an intravascular clot to the lungs. A large embolus may result in sudden death. A smaller clot may produce a localized infarct. This results in dyspnoea, cyanosis, pleuritic chest pain and haemoptysis. Managed:
            • Symptomatically; analgesia with morphine
            • Anticoagulation: with heparin and warfarin
    b Respiratory problems:
        i Collapse of a lung lobe: Treated with deep breathing, postural drainage, adequate analgesia and antibiotics

    ii Aspiration
    iii Infection – bronchopneumonia
    iv Embolus
  c Urinary tract problems:
    i Acute retention of urine
    ii Urinary incontinence
    iii Urinary tract infection
    iv Acute renal failure
  d Gastrointestinal problems:
    i Peritonitis
    ii Paralytic ileus
    iii Mechanical obstruction
  e Pyrexia: Reactionary pyrexia common in the first 48 hours postoperative
  f Anaphylaxis or drug reaction
  g Transfusion reactions

SURGERY ON THE VULVA

Excision (Marsupialization) of Bartholin's cyst

Indications: Cyst in Bartholin's gland
Method:

1 Incision: At the junction of skin and vaginal wall

2 Procedure: cyst wall is separated from adjacent connective tissue. Cyst is removed. The resulting cavity is obliterated with catgut sutures

Excision of urethral caruncle

Method:

1 Caruncle excised with diathermy knife

2 Base of caruncle is then cauterized; a small dilator is passed into the urethral canal to protect its anterior wall

Prolapse of urethral mucous membrane

Method:

1 The stretched prolapsed membrane is transfixed with a catgut suture

2 The prolapsed part of the mucous membrane is removed with scissors

3 The right and left sides of the cut membranes are then approximated

Simple excision of the vulva

Indications
1 Leukoplakia

2 Lichen sclerosis

Method:
1 Incisions:
    a First incision extends around vaginal introitus
    b Outer incision extends from mons veneris outside each labium majus to the perineum

2 Procedure:
    a Skin between the incisions is removed
    b Haemostasis is secured
    c The edges of the two incisions are approximated

Radical excision of vulva

Indications: Carcinoma of the vulva

Method:
1 Incisions:
    a Remove a one-inch strip of skin extending from the anterior superior iliac spine to the pubic spine
    b Incisions around the pudenda for removal of vulva and primary tumour

2 Procedure:
    a Block dissection of lymphatic glands draining the vulva: these are:
        i Superficial and deep inguinal nodes
        ii Cloquet's gland in the femoral canal
        iii Nodes flanking the external iliac vein
    b Removal of vulva by diathermy knife

3 Closure:
   a Upper part of the vulval wound is closed
   b Remainder is allowed to granulate
   c Self-retaining bladder catheter left *in situ* until granulation occurs

Enlargement of the vaginal introitus

Indications: Severe dyspareunia due to:
1 Congenital atresia of the vaginal opening

2 Stenosis following surgery (episiotomy repair)

Method:
1 Incision: T-shaped incision is made:
   a At the border of the posterior vaginal wall
   b Vertically down almost to the anterior wall

2 Closure:
   a Haemostasis secured
   b The cut edge of the dissected vaginal flap is sutured to the skin edge

PROCEDURES ON THE CERVIX

Cervical dilatation

Indications:
1 Preliminary to curettage or other operations

2 Occasionally – for dysmenorrhoea

3 To allow drainage of a pyometra

Procedure:
1 Aseptic procedure

2 Patient placed in lithotomy position. Bladder may be catheterized

3 Bimanual examination of the pelvis

4 Self-retaining speculum (Anvard speculum) is inserted

5 Vulsellum forceps used to grasp the anterior lip of the cervix

6 Uterine sound passed into the uterus to ascertain its position and length

7 Cervix is dilated with graduated metal dilators passed into the uterus

Linear cauterization

Indications:
1 Treatment of cervical erosion
2 Certain cases of chronic cervicitis, where cervical glands are infected

Method: Series of linear applications of cautery probe made beginning in the cervical canal and radiating outwards

Complications: Bleeding may occur due to separation of sloughs on or about the tenth postoperative day

Trachelorrhaphy (cervical repair)

Indications: Operation not often performed. Cervical tear that extends up to the internal os, resulting in an ectropion

Method:
1 V-shaped portion of cervix corresponding to the laceration removed on one or both sides. This should include the apex of the tear

2 The cut edges so produced are united by an interrupted suture

Cervical amputation

Indications:
1 As part of the Manchester operation for uterine prolapse

2 Occasionally:
  a Certain cases of congenital elongation of the cervix
  b Chronic cervicitis with hypertrophy

Method:
  1 A self-retaining speculum is inserted into the vagina. Cervix is drawn down with 2 pairs of Vulsellum forceps

  2 Cervix is then dilated, to facilitate later suturing

  3 Anterior and posterior flaps are made at the appropriate level. These are then reflected upwards

  4 Lateral cervical arteries are tied

  5 Cervix is amputated

  6 The anterior and posterior flaps are sutured to lie over the stump

Complication: Secondary haemorrhage around the tenth day

DILATATION AND CURETTAGE

Indications:
  1 Diagnostic:
   a Cases of abnormal uterine bleeding
   b In cases of amenorrhoea and oligomenorrhoea
   c Investigation of infertility
   d Investigation of congenital malformations
   e Investigation of suspected pelvic tuberculosis
   f Diagnosis of carcinoma of the uterus or cervical canal

  2 Therapeutic:
   a Removal of cervical and endometrial polyps
   b Hypertrophied endometrium
   c Retained products of conception following abortion
   d Termination of pregnancy
   e Drainage of pyometra or haematometra

Procedure:
  1 Dilatation of cervix (see above)

2   Ring forceps are then passed into the uterus to seek polyps

3   Curette is then inserted. It is then drawn from above down-
wards over the internal surface of the uterus

4   Suction evacuation may be used with, or replace curettage,
when:
   a   Terminating a pregnancy in the first trimester
   b   Evacuating an incomplete abortion
   c   Evacuating a hydatidiform mole

NB: In cases of a recent abortion with retained products of conception, a blunt curette, or gloved finger should be used to prevent the risk of uterine perforation

Contraindications:

1   Normal intrauterine pregnancy

2   Acute cervicitis

3   Endometritis

4   Pelvic inflammatory disease

Complications:

1   Cervical laceration from Vulsellum forceps

2   Tear of internal os of the cervix from dilation

3   Cervical incompetence

4   Uterine perforations

OPERATIONS FOR PROLAPSE

Anterior colporrhaphy

Indications:

1   Connection of cystocoele and urethrocoele

2   Stress incontinence due to urethrocoele

3   As part of the Manchester operation (see below)

Method:
1 Incision:
    a Cervix pulled down with Vulsellum forceps
    b Small transverse incision is made in the anterior va-
       ginal wall or cervical-vaginal junction

2 Procedure:
    a The space between vaginal surface and pubocervical
       fascia is exposed, and the vaginal wall bluntly separ-
       ated towards the urethral meatus
    b The separated vaginal wall can now be divided with
       scissors and the flaps reflected laterally
    c The uterovesical ligament is divided and the bladder
       pushed up off the cervix
    d Pubocervical fascia is then pleated with catgut
    e Redundant flap of vagina on both sides is cut away
       with scissors

3 Closure:
    a Vaginal edges approximated with interrupted catgut
       sutures
    b Consider leaving a self-retaining catheter

Colpoperineorrhaphy

Indications:
1 Rectocoele

2 Enterocoele

3 Perineal repair

Method:
1 Incision:
    a Kocher's forceps are fixed to the most lateral caruncu-
       lae hymenales on either side; these mark the points
       where the cut edges of the vagina are approximated
    b The junction between perineum and posterior vaginal
       wall is removed with scissors
    c The rectum having been bluntly separated with
       fingers, a triangular flap of separated vaginal wall
       reaching up to the rectocoele is cut away

2 Procedure: Levatores ani are firmly sutured together in the midline

3 Closure: Cut edges of perineum are reunited with fine catgut

Perineorrhaphy for perineal laceration

Indications: Complete laceration of the perineum

Method:
1 Incision: H-shaped incision. The transverse part of the incision runs along the edge of the fused vaginal walls. The posterior part goes over and beyond the tear of the external superficial sphincter. The anterior limb extends up to the posterior end of the labia minora

2 Procedure and closure:
  a Scarred and adherent vaginal wall, components of the perineal body and anal canal are dissected out. Scar tissue is excised
  b Vaginal wall is dissected up and away from the vaginal canal
  c Anal canal is then dissected away from the scar tissue. Lining of the anal canal is then united with sutures down to the anal margin
  d The internal sphincter, superficial external sphincter, and deep external sphincter are sutured separately
  e Levator ani muscles re-united at their insertion into the deep external anal sphincter
  f Redundant vaginal tissue is removed
  g Perineal body is reconstituted by uniting transversus perinealis muscles and skin

Manchester operation

Indications: Prolapse of the uterus coupled with a cystocoele

Method:
1 Incision:
  a Cervix is pulled down by Vulsellum forceps and three

pairs of marking forceps are applied to the vaginal wall:

   i Just below the urethra

   ii On each side of the cervix at the level of the intended cervical amputation

b Incision is made between forceps (b) and (c) and the triangular piece of the anterior vaginal wall is removed and the cystocoele exposed

c The cystocoele is corrected with an anterior colporrhaphy

2 Procedure:

   a The bladder is pushed off the cervix and the lateral cervical vessels are clamped for ligaturing

   b The cervix is then amputated

   c The posterior vaginal wall is sutured over the posterior lip of the cervix

   d The transverse cervical ligaments are ligatured together in front of the cervix ('Fothergill's suture'). This shortens the transverse cervical ligaments and the uterus thus becomes supported on a higher level

   e The anterior lip of the cervix is covered by suturing over anterior vaginal wall

   f The cut edges of anterior vaginal wall are now approximated

   g The procedure is then completed with a colpoperineorrhaphy

Vaginal hysterectomy and repair of prolapse

Indications:

1 Third degree prolapse of the uterus

2 Prolapse complicated by non-malignant uterine disease

Contraindications:

1 Need to either include an ovariectomy, or perform a laparotomy

2 Mechanical difficulties, such as a narrow vagina or a large or fixed uterus

Method: May be performed in combination with either anterior colporrhaphy or colpoperineorrhaphy

1 Incision:
   a Cervix drawn down by Vulsellum forceps on each lip
   b Incision is made around the cervix down to the space between the vaginal wall and the pubocervical fascia

2 Procedure:
   a Bladder is then freed from front of the cervix and displaced upward, carrying the ureters upward as well
   b The pouch of Douglas is opened. Any excess peritoneum due to enterocoele is removed
   c The uterosacral and transverse cervical ligaments are tied with chromic catgut and divided but held temporarily with artery forceps
   d The uterovesical peritoneum is opened. The uterine arteries are ligatured, divided and held
   e The uterine fundus is now delivered. This exposes the upper part of the broad ligament. The Fallopian tube, round ligaments, and uterine vessels are clamped and cut from the uterus. This frees the uterus
   f All the clamped pedicles are re-tied for safety

3 Closure:
   a Peritoneal edges are sutured together
   b The pedicles are kept outside the peritoneal cavity. The broad ligament pedicles are tied together beneath the bladder. The remaining pedicles are united
   c The vaginal walls are then approximated

Particular points concerning operations for prolapse

Indications for surgery:

1 Symptoms should definitely be due to prolapse. The symptoms may then be temporarily relieved by a pessary

2 The patient should complain of symptoms and if the prolapse is an incidental finding the operation is best deferred

Complications: Urinary, in particular. Urinary tract infection is common, especially if catheterization is performed for acute retention of urine

Late complications include:
1 Recurrence of prolapse:
   a Due to extension of fibrous supports: This may result from increased intra-abdominal pressure (due to obesity or chronic bronchitis) or congenital weakness of the supports
   b Faulty techniques

2 Stress incontinence

3 Dyspareunia: This may result from:
   a Too early resumption of intercourse before full healing
   b Too small a vaginal orifice. Usually stretching occurs. Otherwise surgery to enlarge the vaginal introitus may be needed
   c Too narrow or too short a vagina
   d Adhesions between vaginal walls – not common – usually due to sepsis

4 Complications of cervical amputation:
   a Sterility
   b Premature labour and abortion
   c Precipitate labour
   d Cervical dystocia

OPERATIONS FOR URINARY STRESS INCONTINENCE

Anterior colporrhaphy

Indications: Stress incontinence due to urethrocoele or dislocation of the urethra

Rationale: Corrects prolapse and restores urethrovesical angle which is important in maintaining urinary continence

Method: As above; additionally:
1 Dissection extends all the way to the external urinary meatus

2 Urethra is then separated from vaginal wall. The periurethral connective tissue is then plicated by a suture continuing into the pubocervical fascia

3 Urethra is sutured into the fascia of the triangular ligament and levator ani. This restores the urethrovesical angle

Sling operations: Aldridge's operation

Indication: Failure of anterior colporrhaphy to correct stress incontinence

Rationale: Urethra becomes supported by a sling made from the external oblique aponeurosis. Hence when the rectus muscles contract for example during coughing or sneezing the sling supports the urethrovesical junction and prevents the loss of the urethrovesical angle

Method:
1 Incision:
    a Transverse abdominal incision
    b Two aponeurotic strips from rectus are dissected off, but left attached medially
    c Patient moved into lithotomy position
    d Incision to reach urethra and bladder neck as for anterior colporrhaphy

2 Procedure:
    a Fascial strips then grasped by long forceps introduced lateral to the urethra and pulled into the vagina, on each side
    b Strips are then sutured together close to the bladder neck with enough tension to support the urethra

3 Closure: This is with a self-retaining bladder catheter, left in place for about a week postoperatively

Colpocystopexy: the Marshall–Marchetti operation

Indication: Failure of anterior colporrhaphy for correction of stress incontinence: simpler alternative to Aldridge's operation

Rationale: Urethrovesical junction is firmly fixed to the anterior vaginal wall by suturing the vaginal tissue to the back of the symphysis pubis

Method:
1   Incision: Lower, transverse abdominal incision

2   Procedure:
    a   Urethrovesical junction is exposed in the retropubic space. Urethra is dissected out to reach the external meatus
    b   Foley's catheter is introduced into the bladder. This helps to locate the urethrovesical junction. Vaginal tissue on either side is then sutured onto pubic periosteum
    c   Further sutures may be made between bladder wall and rectus muscle

3   Closure: This is with a suprapubic drain for 48 hours and catheter left in place for 7 days

OPERATIONS ON THE UTERUS

Abdominal myomectomy

Indications: Fibroids in young women of child-bearing age

Contraindications:
1   Relative contraindication: Women over 35–40 years or those who have completed their family

2   Absolute contraindication: Suspicion of malignant disease

Method:
Note: On occasions it may be that myomectomy is impracticable if the number of fibroids is large, or if there is bilateral salpingitis
1   Incision: Low transverse abdominal incision

2   Procedure:

    a Incision in the uterus depends on the position of the fibroid. Incision usually made in anterior part of uterus

    b Uterine cavity opened to exclude submucous fibroids

    c Enucleated cavities closed with sutures

Abdominal hysterectomy

1 Subtotal hysterectomy: The cervix is not removed. It is only justified if total hysterectomy is not feasible

2 Total hysterectomy: Removal of entire uterus, body and cervix. This is the usual method of choice

3 Pan-hysterectomy: Removal of the uterus with both tubes and ovaries

4 Extended hysterectomy: Removal of the uterus, both tubes and ovaries and a cuff of vagina (performed for carcinoma of the uterine body)

5 Radical (Wertheim's) hysterectomy: Removal of the uterus, tubes, ovaries, broad ligaments, parametria, the upper half or two thirds of the vagina and regional lymph nodes (performed for carcinoma of the cervix)

Total hysterectomy

Method:

1 Incision: Low transverse, or subumbilical midline incision

2 Procedure:

    a Broad ligaments, including Fallopian tube, round ligament and ovarian ligament are clamped. Precise position of clamp depends on whether ovaries are to be removed

    b Transverse incision through vesico-uterine peritoneum. Bladder is dissected free from the cervix

    c Parametrium (which contains uterine arteries), and then the uterosacral ligaments, are clamped and divided

    d Having ensured the apex of the vagina is clear of bladder and ureters, it can then be opened to allow excision of cervix and hence uterus. The vagina is then closed

Note: In subtotal hysterectomy the cervix is cut across after being clamped at level of the internal os. This is after clamping the uterine arteries

Particular complications of total hysterectomy:

1   Risk to the ureter if this is not dissected clear of the cervix

2   Psychological postoperative problems owing to loss of child-bearing ability

Radical hysterectomy (Wertheim's)

Indications: Cervical carcinoma; in particular in early cases where the growth is restricted to the cervix. The operation removes the uterus and appendages, and as much cellular tissue of the pelvis and the regional lymphatics as possible

Method:

1   Incision: Transverse abdominal

2   Procedure:
 a   The broad ligament is opened. The ureters are separated from uterus and cervix down to their entry into the bladder
 b   Rectum is then separated from uterus and vagina. Uterosacral and transcervical ligaments are clamped and divided
 c   A Wertheim clamp is placed across the vagina well below the diseased cervix. The vagina is divided below the level of the forceps. The cut edges of the vagina are stitched to prevent oozing
 d   The tissue around the external iliac vessels and in the external obturator fossa are cleared
 e   The iliac lymph nodes and obturator nodes are dissected and removed. Some surgeons begin by a block dissection of the nodes with removal of uterus and cervix as a single unit

3 Closure:
  a The floor of the pelvis is covered by re-uniting the peri-
    toneum
  b Abdomen is closed, possibly with pelvic drainage

Pelvic exenteration

Indications: Certain cervical carcinomas in which growth
involves bladder and/or rectum, not yet involving lateral pelvic
walls. The operations are extensive
Anterior pelvic exenteration:
  1 The uterus is removed as for a Wertheim hysterectomy

  2 Cystectomy with ureteric implantation either into the colon
    (uretero-colic implantation) or ileum as ileal conduit

Posterior pelvic exenteration:
  1 Uterus removed

  2 Lower colon removed with colostomy

Total exenteration:
  1 Uterus, lower colon and bladder removed

  2 This leaves a colostomy and urinary diversion

Particular complications of radical surgery

During surgery:
  1 Haemorrhage

  2 Shock

  3 Damage to nearby organs

  4 Inoperability

Postoperative:
  1 Urinary tract:
    a Ureteric fistulas
    b Complications of urinary diversions:
      i Hyperchloraemic acidosis
      ii Ascending infection

    c Urinary retention

2 Gastrointestinal tract:
    a Paralytic ileus
    b Intestinal obstruction

3 Infection

4 Thromboembolic complications

5 Lymphocoele formation

OPERATIONS ON THE OVARIES

Oophorectomy

Removal of the ovary:
Indications:
1 Normal ovaries when performing total hysterectomy in women approaching the menopause

2 Hysterectomy for malignant disease

3 Malignant disease in ovaries

4 Ovarian cysts

Ovarian cystectomy: Enucleation of a benign tumour or cyst from its capsule with preservation of the ovary

Indications: Ovarian cysts, including dermoid cysts, serous cysts and fibromata

Method:
1 Capsule of cyst is lightly incised

2 Cyst enucleated

3 Ovary is reconstituted by fine catgut sutures

Decortication or wedge resection: Performed for sclerocystic disease of the ovary and in the Stein–Leventhal syndrome

OPERATIONS ON THE FALLOPIAN TUBES

Surgery for tubal occlusion

Salpingostomy

Indications: To open tubes closed by infection

Rationale: Formation of a new abdominal opening to the Fallopian tube

Method:

1 A crossed incision is made at the abdominal end of the ampulatory portion of the tube

2 Its mucous membrane is then everted and stitched in this position

Tubal implantation

Indication:

1 Occlusion in the cornua

2 Occlusion near the uterine end (isthmus) of the Fallopian tube

Method:

1 Abdomen is opened

2 Tube is insufflated from the distal end to determine the position of the obstruction

3 Tube is transected distal to the obstruction, and its cut end sutured to the uterus such that their lumina communicate

4 Patency is ensured by 'splinting' with polythene tubing at the anastomosis for several weeks postoperatively

Salpingectomy

Definition: Removal of the Fallopian tube

Indications:

1 Disease of the tubes:

   a Salpingitis
   b Pyosalpinx
   c Tubal pregnancy

 2 Sterilization (partial salpingectomy)

Method:
 1 Incision: Abdominal
 2 Procedure:
   a The Fallopian tube is mobilized and freed and then removed between forceps
   b Haemostasis is secured and the abdomen closed

Sterilization

Note: Medico-legal aspects of sterilization are discussed in Chapter 8

Indications:
 1 Presence of any illness that may endanger life, or endanger health during pregnancy

 2 As a method of contraception

Methods:
 1 Involving abdominal incision: Loop of tube is tied and the tube enclosed by the ligature is excised

 2 Involving laparoscopy: Tubes are visualized and can be
   a Cauterized
   b Clipped with Hulka clips, or
   c Ringed with Falope rings

Index

218